MATLAB®

WITH APPLICATIONS
TO ENGINEERING,
PHYSICS AND
FINANCE

MATLAB®

WITH APPLICATIONS TO ENGINEERING, PHYSICS AND FINANCE

DAVID BAEZ-LOPEZ

CRC Press
Taylor & Francis Group
Boca Raton London New York

CRC Press is an imprint of the
Taylor & Francis Group, an **informa** business

A CHAPMAN & HALL BOOK

CRC Press
Taylor & Francis Group
6000 Broken Sound Parkway NW, Suite 300
Boca Raton, FL 33487-2742

First issued in paperback 2019

© 2010 by Taylor & Francis Group, LLC
CRC Press is an imprint of Taylor & Francis Group, an Informa business

No claim to original U.S. Government works

ISBN-13: 978-1-4398-0697-5 (hbk)
ISBN-13: 978-0-367-38498-2 (pbk)

Library of Congress Cataloging-in-Publication Data

Báez López, David.
 MATLAB with applications to engineering, physics and finance / David Baez-Lopez.
  p. cm.
 Includes bibliographical references and index.
 ISBN 978-1-4398-0697-5 (hardcover : alk. paper)
 1. Numerical analysis--Data processing. 2. MATLAB. I. Title.

 QA297.B28 2010
 620.001'51--dc22   2009031934

Visit the Taylor & Francis Web site at
http://www.taylorandfrancis.com

and the CRC Press Web site at
http://www.crcpress.com

"i carry your heart with me (i carry it in
my heart) i am never without it (anywhere
i go you go, my dear; and whatever is done
by only me is your doing, my darling)
i fear
no fate (for you are my fate, my sweet) i want
no world (for beautiful you are my world, my true)
and it's you are whatever a moon has always meant
and whatever a sun will always sing is you
here is the deepest secret nobody knows..."

e. e. cummings

Dedicated to my children:

Laura Michele

and

David Alfredo "Gary"

Table of Contents

Preface

Mathematics is used in almost every field of knowledge. For example, it is necessary in engineering and sciences, finance, biology, chemistry, and accounting, to name a few. Mathematics is taught at all levels of education, from elementary to college and graduate school. Thus, most people have a fairly good knowledge of some area in mathematics. Unfortunately, most students and mathematics users are not taught using mathematics software tools like MATLAB®[1], among other similar tools such as Mathematica and MathCAD, which allow users to solve mathematics problems when they arise in their corresponding fields of expertise. This is the purpose of this book: to allow mathematics users to learn and master the mathematics software package MATLAB.

MATLAB integrates computation, visualization, and programming to produce a powerful tool for a number of different tasks in mathematics. MATLAB is the acronym for MATrix LABoratory. In MATLAB, every mathematical variable or quantity is treated in matrix form.

With MATLAB we can perform complex mathematical tasks with relatively simple programs. This is possible because MATLAB has close to 10,000 built-in functions from simple ones such as differentiation, integration, and plotting, to optimization functions which require no user programming.

Many of the functions mentioned above are grouped into toolboxes especially dedicated to some field of science, finance, or engineering.

Simulink® is another software package that runs from MATLAB. Simulink simulates systems at the block level. Thus, it is ideal for scientific and engineering system simulation. Simulink is described in Chapter 8 and some examples show the great advantages of using it for system modeling and simulation.

There are many books available on MATLAB, but a unique feature of this book is that it can be used by novices and experienced users alike. It is written for the first-time MATLAB user who wants to learn the basics of MATLAB, but at the same time, it can be used by users with a basic MATLAB knowledge who want to learn advanced topics such as programming, create executables, publish results directly from MATLAB programs, and create graphical user interfaces. In addition, for experienced users, it has three chapters with MATLAB applications in engineering, physics, and finance. It also includes sets of exercises at the end of each chapter. Each and every one of the examples and exercises were solved using MATLAB 7.8.0 Release 2009a, and a solutions manual is available from Taylor & Francis Group, LLC.

---

[1]For product information, please contact: The MathWorks, Inc., 3 Apple Hill Drive, Natick, MA 01760-2098, USA, Tel: 508-647-7000, Fax: 508-647-7001, E-mail:info@mathworks.com, Web: www.mathworks.com.

The author wishes to thank those in the Book program from The Math-Works, Inc., especially the help of Naomi Fernandes. The author also wishes to thank the staff at Taylor & Francis: Li Ming Leong, Marsha Pronin, and Michele Dimont. Finally, I give thanks to my son David Alfredo Báez Villegas, from Instituto Nacional de Antropología e Historia in México City, for reviewing the complete manuscript, and to my student Oswaldo Cruz-Corona for having helped me to obtain some of the figures.

David Baez-Lopez
Department of Computers, Electronics, and Mechatronics
Universidad de las Américas
Cholula, México

Chapter 1

Introduction to MATLAB

1.1 Introduction

MATLAB® is a very high-level powerful system designed for technical computing. It integrates in the same software environment computation, plotting, and programming. It also has a very easy mathematical notation. Some of the most common applications for technical computing are as varied as:

Technical computing
Algorithmic development
Modelling and simulation
Data analysis
Plotting
Graphical User Interfaces

MATLAB is an acronym for **MAT**rix **LAB**oratory and it was originally developed to perform matrix calculations. MATLAB programming language is, probably, more powerful than traditional programming languages such as C, C++, VisualBasic, to name a few.

MATLAB was developed in 1984 by Cleve Moler and Jack Little, who founded The MathWorks, Inc. in Natick, Massachusetts. The first version of MATLAB only had about eighty functions. The very latest version includes more than ten thousand functions.

Besides MATLAB, The MathWorks has developed a series of software packages, called toolboxes, written in the MATLAB programming language. These toolboxes can perform a number of calculations in several branches of engineering, economics, finance, physics, and mathematics, among others. It is hard to imagine an area of knowledge where MATLAB does not have an application. In the coming chapters we will use examples of the different areas where MATLAB has applications to illustrate how to use MATLAB.

1

Figure 1.1: Cleve Moler is president and principal investigator in The MathWorks, Inc. In a 20-year span he was a professor at the University of Michigan, Stanford University, and the University of New Mexico.

Figure 1.2: Jack Little holds a B.S. degree from MIT and an M.S. from Stanford University (1980), both in Electrical Engineering. He is a fellow of the IEEE and has worked in the Control and Signal Processing toolboxes.

1.1.1 Book Organization

The book is organized in the following way. The first two chapters are an introduction to MATLAB. The following three chapters, 2 to 5, cover basic calculation in MATLAB. The topics include linear algebra, calculus, and plotting. The next three chapters, 6 to 8 cover programming, advanced programming techniques, graphical user interface (GUI) development, and Simulink which is a MATLAB-based GUI useful for system modelling and simulation. The last three chapters cover a broad set of examples illustrating applications in several engineering disciplines, physics, and finance.

Each chapter contains a set of end-of-chapter exercises. These exercises complement the material covered in the chapter and are an important part of the chapter. It also contains a glossary of terms explaining those that could be known only to experts in the field but which are useful to know for a better understanding of the examples and exercises.

1.1.2 Chapter Organization

The chapter is organized as follows. It begins with a MATLAB environment description. It describes the different windows available. It continues with basic calculation describing how to create and store variables. Different formats for variables are introduced. Basic functions are presented and the set of elementary function is described. Plots are created using simple functions. This topic is further explored in Chapter 5 but since plots are used in almost any chapter, here we give a brief introduction to it. A special set of variables called strings are introduced and some operations on strings are described. A very useful characteristic of MATLAB is the fact that all the variables, functions, and calculations are stored in the Command History and thus, it can be stored in a file, or a program can be created from them. This is covered in detail in this chapter. Finally, it is explained how to use MATLAB Help.

1.2 Starting MATLAB

To install MATLAB, follow the instructions provided by The MathWorks, Inc. to install the software and the licenses. When the installation is finished, a MATLAB icon will de displayed on the Desktop and in the Programs path. When we click on the icon, MATLAB will open, as shown in Figure 1.3. In the main menu in the Desktop we can display up to four Windows: Command Window, Workspace, Current Directory, and Command History. Each of these windows can be attached to the main window, but we can have any of them as a separate window by using the right upper arrow. We can reattach the window using the same arrow. The Command window is the window where we enter variable values and instructions to perform calculations. The Workspace contains the variables created in the current session. The Current Directory displays the files and folders in the current directory. The Command History

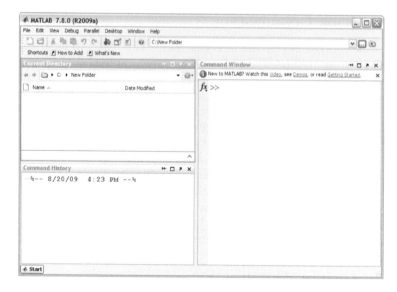

Figure 1.3: MATLAB main window.

displays the instruction executed in the current and previous session, unless it has been deleted.

1.3   Simple Calculations in MATLAB

MATLAB can perform simple calculations as if it were a simple calculator. For example, if we wish to add two numbers such as 2+3, we simply write in the Command Window

>> **2+3**

and press **ENTER**, the output that MATLAB gives is

ans=
    5

We use the convention that the data we enter in MATLAB is written in bold and the results from MATLAB are written in regular type. Also note that the text is written in Times Roman font and MATLAB input and output are in a SansSerif font.

We now show in Table 1.1 the basic operations and some elementary functions available in MATLAB. The elementary functions are available in the set of functions called **elementary functions** group.

Table 1.1: Basic MATLAB operations and their precedence

Operation	Symbol	Example	Precedence
Addition	+	15+8=23	3
Subtraction	-	16-11=5	3
Multiplication	*	3*7.2=21.6	2
Division	/	18/3=6	2
Exponentiation	^	3^4 =81	1

The precedence means the priority order to perform the different operations. So, exponentiation has the highest priority, multiplication and division have the following higher priority, and finally, addition and subtraction have the lowest priority. Thus, for example, in the operation

>> **2*3+4*7^4**

The exponentiation 7^4 is performed first, then the multiplication 2*3 and 4 times the result of 7^4, and finally, the addition of the results of 2*3 and 4*7^4.

Precedence can be user modified by using parentheses. Operations enclosed by parentheses are performed first. For example,

>> **2^3*4**

gives the result

ans=
      32

Since the first operation is 2^3 which gives 8 and then this result is multiplied by 4. Instead, we can use parentheses as in

>> **2^(3*4)**

and we obtain

ans=
      4096

because first MATLAB performs (3*4) which gives 12 and then 2^12 which gives 4096.

Table 1.2: Elementary functions in MATLAB

Function	MATLAB Notation
$\sin x$	sin (x)
$\cos x$	cos (x)
$\tan x$	tan (x)
$\sqrt{x}$	sqrt (x)
$\log_{10}(x)$	log10 (x)
$\ln(x)$	log (x)
$|x|$	abs (x)
e^x	exp (x)

1.3.1 Elementary functions

MATLAB has a set of functions of almost everyday use. These functions are available in a set called Elementary Functions. The trigonometric functions fall within this set. For example, to calculate sin(1)

>> **sin(1)**

ans=
     0.8415

For the case of the trigonometric functions, the argument is in radians. Some of the elementary functions are available in Table 1.2. For a listing of all of the elementary functions, we enter help elfcn in the Command Window. elfcn is an acronym for **elementary functions**. As mentioned above, MATLAB has more than ten thousand functions. Some of the most elementary ones are written in C and most of them are written in the MATLAB language.

For the case of trigonometric functions, a simple example is to find the sine of the irrational number π. If we approximate π by 3.1416, then

>> **sin (3.1416)**

ans=
     -7.3464e-006

which is a good approximation to the exact result. MATLAB has a predefined value for the constant π and called simply as pi. Then, we have

>> **sin (pi)**

ans=
     1.2246e-016

which is closer to the expected result. Other examples are

>> **sqrt (2)**

ans=
    1.4142

>> **log10 (1000)**

ans= 3.0000

Some of the predefined constants are:

MATLAB	
pi	3.14159265
i	Imaginary unit $=\sqrt{-1}$.
j	Same as i.
eps	This is the precision of the floating point operations, $2.22*10^{-16}$
Inf	Infinity
NaN	Not a number.

1.4 Variables

Variables are created in MATLAB as they are defined. That is, there is no need to predefine them as it has to be done in languages such as C, Fortran, and Visual Basic, to name a few. For example, the variable alpha is created the first time we used as:

>> **alpha=32**

alpha=
    32

Now, this variable appears in the Workspace window and will remain there until we delete it with the instruction clear which erases all variables stored in the workspace. To see a list of variables in the Workspace we only need to type who. For example, if the session just started with this section, the only variable is alpha. Then, we obtain

>> **who**

Your variables are: alpha

Table 1.3: Formats for displaying numerical values

MATLAB format	Displayed value	Comments
format short	1.4142	5 digits
format	1.4142	Same as format short
format long	1.414213562373095	16 digits
format short e	1.4142e+000	5 digits plus exponent
format long e	1.414213562373095e+000	16 digits plus exponent
format hex	3ff6a09e667f3bcd	hexadecimal
format bank	1.14	2 decimals (for currency)
format +	+	positive or negative
format rat	1393/985	Approximates to a ratio

Variable names can have up to 63 characters. If a variable name is longer than 63 characters, the name will be truncated to 63 characters.

The number of digits used by MATLAB can be changed according to the format chosen for this purpose. The formats available are shown in Table 1.3. We show the formats with the irrational number $\sqrt{2}$.

Each time MATLAB performs an action, the result is displayed in the Command Window. Sometimes we are only interested in the final results and not in the intermediate ones. We can suppress intermediate results by typing a semicolon after the line. For example:

```
>> 23+32

ans=
   55

>>
   23+32;
```

The first time we calculate 23+32 we get an answer, the second time we have a semicolon after the indicated addition and we **do not** get an answer at all; however, the result is stored in the variable ans.

MATLAB variable names are case sensitive, so care must be given when writing long programs as we see in Chapter 6. Thus, variable A1 is different from variable a1.

For a set of numbers written between brackets as in

```
v=[ pi 2 -4 8 sin(2)]
```

The variable v is called a vector, in this case a row vector. In a column vector the elements are separated by semicolons, thus it is written as w=[2 ; 3; -7; sqrt(3)]

In column vector w the elements are separated by a semicolon. This indicates that the elements after the semicolon are located in a different row. Thus, MATLAB gives as an answer to these two vectors:

>> v=[pi 2 -4 8 sin(2)]

v =
    3.1416 2.0000 -4.0000 8.0000 0.9093

>> w=[2 ; 3; -7; sqrt(3)]

w =
2.0000
3.0000
-7.0000
1.7321

In the case of row vector v, the elements can be separated by either blanks or by commas while in a column vector, elements are separated by semicolons. A vector has dimension n if it has n elements. Thus, vector v has dimension 5 whereas vector w has dimension 4.

If we have two vectors with the same dimension, we define the dot product by

$$x'*y$$

The result is a scalar. Thus, for vectors v, w given by

>>v=[-1 4 7 -9]

v =
    -1 4 7 -9

>>w=[9 8 -6 3]

w =
    9 8 -6 3

>>v*w'

ans =
    -46

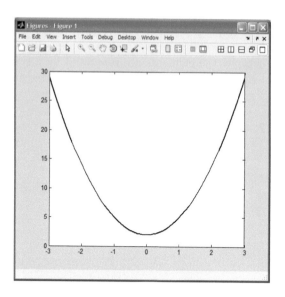

Figure 1.4: Plot of $3x^2 + 2$.

1.5   Plotting

Plotting a function in MATLAB is a very easy task. For example, to plot
$y = 3x^2 + 2$, in the rank [-3, 3] we need to create a set of points x where the
function is to be evaluated. Then, simply use the instruction plot(x,y). We
do this as shown below:

>> x= -3: 0.1: 3; % The set of points for variable x is created.
>> %This creates a vector of values -3,-2.9, -2.8...,2.9,3.
>> y= 3*x.^2+2;% The function is evaluated at the x points.
>> % Here the function is evaluated at each point in the set x.
>> plot (x,y) % Here the function is plotted.

The first row has the format

x = initial point : increment : final point

The second row defines the function we wish to plot. The third row calculates
the function at the points x. We have used the instruction x.^2 which indicates
the operation x^2 is done to each term of the vector x. That is

x.^2=[(-3)^2 (-2.9)^2 ... 2.9^2 3^2]

The last row does the plotting with the instruction plot. The curve is shown
in Figure 1.4.

We can add a title and text to the plot just obtained. We do this with the following instructions:

```
>> xlabel ('plot of 3x^2+2')
>> title ('Plot of a parabole','FontSize',12)
```

In the Edit menu in the plot window, we can open different menus to edit the plot properties, including the axes, and any object selected from the plot such as grid, trace, axis, title color, etc. If we wish to plot more than a trace in the same plot, we can do it in two different ways. For example, if we wish to plot the function z=-0.4*x^3-1 in the same figure we can use the following instructions

```
>> hold on
>> z=-0.4*x.^3-1;
>> plot(x,z)
>> xlabel ('plots of 3x^+2 and -0.4x^3-1')
>> title ('Plot of a parabole and a cubic function','FontSize',12)
```

which changes Figure 1.4 to Figure 1.5. The instruction hold on indicates that we are going to use the same figure where the previous function was plotted. If we continue to use plot instructions, the graphs are going to be plotted in the same figure unless we write hold off. Another approach is to plot both functions with a single plot as in

```
>> plot (x, y, x, z)
>> legend ('parabole', 'cubic')
```

which produces the plot in Figure 1.6. The legend instruction is used to identify the traces in the plot.

It is possible to produce three-dimensional plots. Such plots are called surfaces. For example, (we have omitted the >>)

```
% First we create a grid where we evaluate the function.}
[x,y] = meshgrid (-5: 0.2: 5);
% Evaluate an intermediate variable.
R=(x.^2+y.^2).^0.5;
% Evaluate the z variable.
z=cos(R);
% Finally, create the plot.
mesh(x,y,z);
```

The resulting plot is shown in Figure 1.7. Additionally, the figure's tool bar has an icon to change the observation point. For example, in Figure 1.8 we have changed the observation point. Chapter 5 is dedicated to plotting

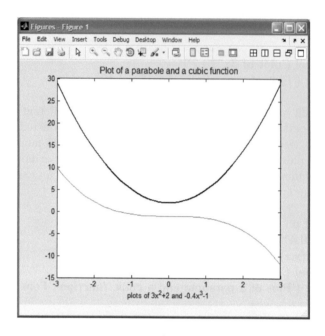

Figure 1.5: Plots of $3x^2 + 2$ and $z = -0.4x^3 - 1$.

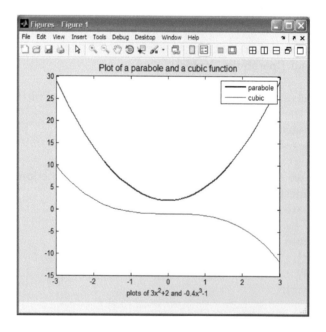

Figure 1.6: Multiple plots with **legend**.

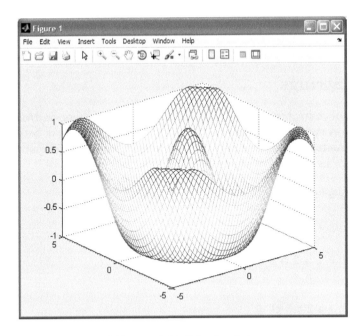

Figure 1.7: Mesh plot.

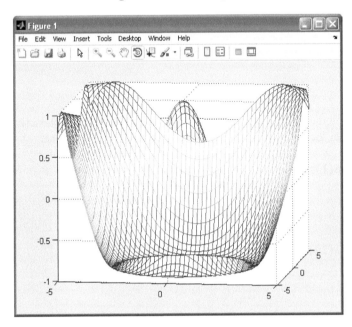

Figure 1.8: Mesh plot with a different observation point.

instructions and we give an extensive treatment for two- and three-dimensional plots.

1.6 Strings

A string of characters is a series of letters, and/or numbers, and/or any other symbols. A string is defined by enclosing the set of characters between single quotes. As examples of strings we have 'book', 'technical computing', '2332'. A variable in MATLAB can be a string. For example,

>> a= 'taylor'

a = taylor

>> b='1 a 2 b 3 c!"#'

b =
    1 a 2 b 3 c!"#

If for any reason a string has to have a quote as a part of the string, we simply repeat the quote as in

>> a='Laura"s bear'

a =
    Laura's bear

The elements in a string have a position. Thus $a(k)$ is the kth element in the string. For the string a given above, $a(1)=L$ and $a(10)=e$,

>> a(1)

ans =
    L

>> a(10)

ans =
    e

To specify a range of characters we use $a(k:n)$ which indicates the characters from the kth to the nth. For example,

>> a(3:9)

ans =

   ura's b

We can also use a(3:2:9) which indicates the set of characters in a string from index 3 to index 9 but incrementing the index by 2, that is, we are looking for a(3), a(5), a(7) and a(9),

>> **a(3:2:9)**

ans =

   uasb

We can concatenate several strings by using

New_string = [string 1, string 2, string 3]

For example, if we have three strings a1, a2, and a3, defined by

>> **a1='Canada '; a2='Mexico '; a3='USA ';**

We can form

>> **b=[a1,a2]**

b =

   Canada Mexico

>>**b2=[a1, a3]**

b2 =
   Canada USA

We can also form new strings with elements from the strings already defined as

>>**y1 = [a1(1:3), a2(1:3), a3(1:2)]**

y1 =
   CanMexUS

There exists a set of functions that can be performed on strings. Here is a list of some of the most used ones:

MATLAB	Function
length	Number of characters in a string.
strcmp	Compares two strings.
str2num	Converts a string to a numerical value.
num2str	Converts a number to a string.
strrep	Replaces characters in a string with different characters.
upper	Changes lowercase characters to uppercase.
lower	Changes uppercase characters to lowercase.

For the function length and the string a='This is Chapter 1', we have

>>length(a)

ans =
    17

The instruction strcmp(a , b) compares two strings a and b. If they are equal the result is unity, but if they are not equal the result is a zero. For example, for the strings a1 and a2 given above,

>> strcmp(a1, a2)

ans =
0

>> strcmp(a2,'Mexico ')

ans =
1

The instruction str2num converts a string to a number only if the string characters are numbers, as

>> d=str2num('2010')

d =
2010

The instruction num2str transforms a number into a string. For the number a=810

>> **num2str(a)**

ans =
   810

The instruction strrep (c1 , c2 , c3) replaces the string c2 by string c3 in the string c1. For example, in the string 'MATLAB is a high level programming language' we can replace the string 'level' by 'level technical' with

c1 = 'MATLAB is a high level programming language' ;
c2 = 'level';
c3 = 'level technical';
strrep(c1, c2, c3)

ans =
   MATLAB is a high level technical programming language

The instructions upper and lower convert lowercase characters to uppercase and vice versa, respectively. For the strings a = 'American Continent' and b = 'European Continent' we have:

h = upper(a)
k = upper(b)
lower(h)
lower(k)

h = AMERICAN CONTINENT
k = EUROPEAN CONTINENT
ans = american continent
ans = european continent

1.7 Saving a Session and Its Variables

A variable and its value are kept only during the session and only until the user uses the instruction clear which deletes all the variables in the session. However, in several cases it is desired to have them available in another session. In other cases, we wish to use not only to keep the variables, but also the instructions executed during the session. That is, we might need to keep the complete session. This can be done using the instruction diary followed by the file name. The instructions and variables will be stored in that file until the instruction diary off is executed, as in the following, diary file_name

   ⋮

diary off

where the dots indicate the instructions and variables. Let us consider the following session

```
>>a=1; b=2;
>> diary session.txt
>> b=3;
>>a=b^2

a =
   9.0000

>>x=-b/a

x =
   -0.3333

>> diary off
>> d=3*x

d =
   -1.0000
```

Now we take a look at the Current Directory and we observe that there is a new file called session.txt. If we open it with File→Open we see that all the variables and instructions together with the responses executed after diary session.txt and before diary off are stored in the file. This is shown in Figure 1.9. If a file name is not specified, the default file name is diary.m.

Another way to save information is by using the command Save Workspace As which is available from File→Save Workspace As or with File→Save. In this case we **only** save variable information. We cannot save instructions with this option. In this case the file name has the extension .m.

These variables are now saved in the file variables.mat. Alternatively, we can also execute

save('variables')

To load the variables in a later session we simply execute

load ('variables')

Figure 1.9: Instructions and variables saved with diary.

We can also use File→Open and select the file variables.

From the Command History window we can also save information. By selecting with the left button mouse the instructions and variables and then with a right click we select the option Create M-File as shown in Figure 1.10. The file name is saved with the extension .m. The final file is shown in Figure 1.11.

1.8 Input/Output Instructions

Up to now we have used MATLAB as a calculator, where the result is written right after we hit the ENTER key. Sometimes this is not very convenient, especially if we are not interested in intermediate results but rather at the final one. Fortunately, we can omit intermediate outputs by typing a semicolon after the instruction, as we saw in Section 1.4.

1.8.1 Formatted Output

Besides the direct output obtained after entering an instruction and hitting the ENTER key, we can use the instruction fprintf. We illustrate its use with an example. To write the string 'This is chapter 1.' we use:

>> fprintf (' This is chapter 1.\n');

This is chapter 1.

We note several things. First we added \n which indicates that there is

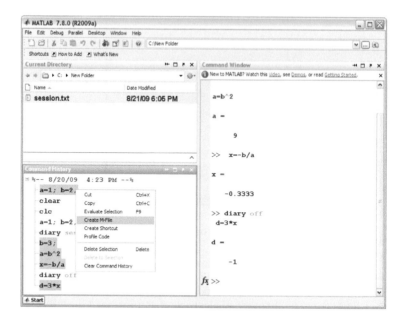

Figure 1.10: Selecting instructions and variables in the Command History window using the mouse right button.

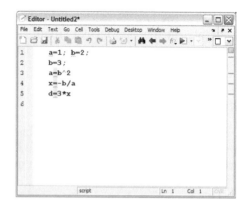

Figure 1.11: Instructions and variables saved in the file **variables.m**.

a line feed after writing the desired string. If we now use again the fprintf instruction, the string included there will be written in the following line. Second, the semicolon does not preclude writing the desired output. And third, there is not a line with ans as before. If we continue writing data with fprintf but we do not use \n, the output is going to be written in the same line. This is so because the instruction does not make a line feed by itself, but we have to indicate it. For example,

>> **fprintf('This '); fprintf('is '); fprintf('Chapter 1.');**

This is Chapter 1.

>> fprintf('This \n'); fprintf('is \n'); fprintf('Chapter 1.\n');

This
is
Chapter 1.

The first time we executed the fprintf instruction, the strings are written in the same row because we omitted the line feed part. In the second fprintf instruction we placed an n\ before ending each string. Thus, after each fprintf instruction there is a line feed.

We can format the output variables. We have different formats to do this. The formats are:

f - floating point
g - fixed or floating point
i - integer
c - character
s - string
d - double precision
e - exponential notation

We can also use fprintf with numerical quantities. For example, for the number a=pi, we can use

>> **fprintf ('The number is %12.8f\n', pi)**

The number is 3.14159265

which means that the number pi is written as a fixed point number with 8 decimal places.

1.8.2 Data Input

The easiest way to input data to MATLAB is through the keyboard, in the same way we have done it so far. Another way is to use the instruction input which has the format

>   x = input ('Enter the value of x')

This instruction displays the string between quotes and waits for the value of x. So we can have the following:

>   >> x = input ('Enter the value of x ') ;

Enter the value of x **56**

>   >> fprintf ('You entered the number %g.\n ', x)

You entered the number 56.

The instruction input does not generate a line feed. As mentioned before, the line feed can only be generated with \n, which can go anywhere within the string.

If we enter an alphanumeric value instead of a numerical one, we get an error message as in:

>   >> x = input ('Enter the value of x ')

Enter the value of x

fifty six

??? Error: Unexpected MATLAB expression.

Enter the value of x

Here we entered a text **fifty six** and get back an error message. MATLAB requests again the value of x. It keeps doing the same thing until we enter a numerical value.

To enter a string we add 's', as in:

>   >> x = input ('Enter the value of x ','s')

    Enter the value of x

fifty six

x =
fifty six

Since every input can be seen as a string, we never get an error message in this case.

1.9 Help

Help in MATLAB can be retrieved from the Command Window by typing help to obtain a list of topics.

>> **help**

My Documents	- (No table of contents file)
matlab\general	- General purpose commands.
matlab\ops	- Operators and special characters.
matlab\lang	- Programming language constructs.
matlab\elmat	- Elementary matrices and matrix manipulation.
matlab\elfun	- Elementary math functions.
matlab\specfun	- Specialized math functions.
matlab\matfun	- Matrix functions - numerical linear algebra.
matlab\datafun	- Data analysis and Fourier transforms.
matlab\polyfun	- Interpolation and polynomials.
matlab\funfun	- Function functions and ODE solvers.
matlab\sparfun	- Sparse matrices.
matlab\scribe	- Annotation and Plot Editing.
.	
.	
.	

For more help on directory/topic, type "help topic".
(Only the first few lines are displayed here).

To obtain help for a particular subject, type help followed by the function name. For example, to obtain information about the elementary functions we type help elfun to obtain a list of all the functions in the set elfun.

>> **help elfun**

To obtain information about a specific function, we type help followed by the function name. If we wish to get information about the sine function we use

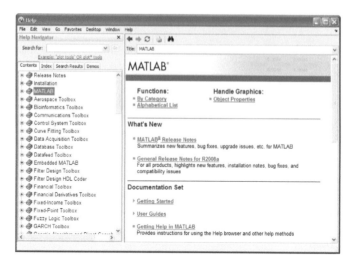

Figure 1.12: Help window.

>> **help sin**

SIN Sine of argument in radians. SIN(X) is the sine of the elements of X.
See also asin, sind.

1.9.1   Help Page

We can also get help from the Help window. This window is opened by
clicking on the Help icon in the Tools bar. We can also access this page by
typing help in the Command Window. After doing this we obtain the window
shown in Figure 1.12. In this window we can see two parts: the left-hand side
which contains the topics and the search part and the right-hand side which
contains the information for the topic selected. The left-hand side has four
tabs: Contents, Index, Search Results and Demos. The Search Results tab has
a field for data labeled Search for. There we write the instruction name or
part of it to look for every instruction which has that name. The information
is displayed in the right-hand window. Figure 1.13 shows an example for the
function diff which calculates the symbolic derivative of a function.

1.10   Concluding Remarks

We have started using MATLAB in this chapter. In the following chapters
we will cover how it can be used to perform an unlimited number of tasks,
going from the very obvious ones such as an addition to very complicated ones
such as optimization of a product or process. MATLAB is used in education,

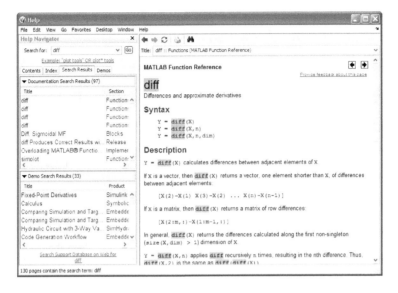

Figure 1.13: Search results for the function **diff**.

research, and testing to name a few applications. We have only described the environment where all those calculations can take place.

In this chapter, we only have covered the way the several types of variables are defined and stored, both numeric ones and strings. Strings are a type of non-numeric variable which finds use in the creation of text for input and output information and for names of variables. We have also described the way to format data for input and output instructions.

Finally, we covered the way a session, including variables and instructions, can be saved for later use.

1.11 Exercises

Section 1.1

1.1 Simultaneously display the main four MATLAB Windows: the Command Windows, the Workspace, the Current Directory, and the Command History.
1.2 Using the arrow in the upper-right corner in the Command History. Using the same arrow, which now has changed direction, return it to its original position.

Section 1.2

1.3 Evaluate the function log(10) and the function log10(10). Explain the difference between both logarithms.

1.4 For x defined by por x =[1 2 3] evaluate p(x) = 2*x+3.

Section 1.3

1.5 Using the formats long and short evaluate the square root of 2.
1.6 Using the formats rat y long evaluate sin(2).
1.7 Find the values of 4, 16, 66 in hexadecimal format.
1.8 Use the instruction who to see the variables in the Workspace.

Section 1.4

1.9 Plot the function exp(x^2) for values of x between -2 and 2.
1.10 Plot the function f(x) = x^2*sin(x) for values of x between 0 and 10. Add an adequate title to the plot. Also add names to the axes.
1.11 Plot the functions f(x) = 1/x and g(x) = 1+sin(x)cos(x) from x=1 to x=2. Add a legend instruction to the plot.
1.12 Make a mesh plot for the function f(x,y) = (y^2)*sin(x*y). Variables (x, y) both are in the range -3 y +3.

Section 1.5

1.13 Generate the strings: "MATLAB is a technical tool" and "powerful". Using length, find out how many characters each string has.
1.14 Using the strings from Exercise 1.13, do a concatenation to form a single string.
1.15 In the strings from Exercise 1.13, add the word very to the string powerful and concatenate the result to form a single string.

Section 1.6

1.16 Perform the following computations in the Command Window:

```
x=pi;
y=2
z=sin(x)
w=cos(x)*z
```

(a) Save the session using Create M-File.
(b) Save the variables in a file using diary.

1.12 References

[1] A. Gilat, MATLAB: An Introduction with Applications, J. Wiley and Sons, New York, 2008.

[2] D. C. Hanselman, Mastering MATLAB 7, Prentice Hall, Inc., Piscataway, NJ, 2004.

[3] R. Pratap, Getting Started with MATLAB 7: A Quick Introduction for Scientists and Engineers, Oxford University Press, New York, 2005.

[4] B. Hahn and D. Valentine, Essential MATLAB for Engineers and Scientists, Third Edition, Butterworth-Heinemann, Burlington, MA, 2007.

[5] H. Moore, MATLAB for Engineers, 2nd Edition, Prentice Hall, Inc., Piscataway, NJ, 2008.

Chapter 2

Variables and Functions

2.1 Variables

MATLAB has a very powerful way to define variables. Most programming languages have to define variables before they are used for the first time. MATLAB does not need to do that. Each variable is defined as it is used. If in a procedure we need to define the variable Y0, it is defined the first time we use it. Thus, if we write

$>>$ **Y0 = 1**

Y0 $=$
   1

This variable has that value until we change its value or we delete it with the instruction clear Y0.

A variable name can have up to 63 characters. If a variable name is longer than 63 characters, MATLAB keeps only the first 63 characters and eliminates the remaining ones.

Variable names have to start with a letter and it can only have letters, numbers and underscore.

All the variables are real unless we define them as complex. The variable Y0 defined above is a real variable. To define a complex variable, we use the imaginary number $i = j = \sqrt{-1}$. We can define a complex number as:

$>>$ **Z = 2+3*j**

Z $=$

   2.0000 + 3.0000i

In the case of complex numbers such as $c = a+b*i$, a and b are called the real and imaginary parts, respectively. They can be found with real(c) and imag(c). For the complex number z we have

```
>> real(z)
ans =
    2
>> imag(z)
ans =
    3
```

ans is a variable that MATLAB creates to give the result of a calculation when we do not assign a variable to the results of imag(z) or real(z).

2.1.1   Symbolic Variables

A symbolic variable is a variable that does not have a numeric value. We have to have installed the Symbolic Math Toolbox. This toolbox is necessary to define and use symbolic variables and functions. A symbolic variable can be defined with

$$a = sym('a')$$
$$x = sym('x')$$

or simply

$$syms\ a\ x$$

As with numeric variables, a symbolic variable starts with a letter. A symbolic variable can be neither i nor j.

If a and x are symbolic variables, we can define symbolic functions. For example, we define the function $f(t)$ as

$$f(t) = 10\sin(3\omega t + \theta) \tag{2.1}$$

In MATLAB we use:

```
>> syms w t theta
>> f = sym('10*sin(3*w*t+theta)')
f =
    10*sin(3*w*t+theta)
```

To define a symbolic complex variable we can first define the real and imaginary parts as real variables and then define the new variable as complex. For example, if

$$z = x + iy \tag{2.2}$$

Then, the complex variable z is defined as

```
>> syms x y real
>> z = x+i*y
z =
    x+i*y
```

with x, y, z we now can realize calculations taking them as either real or complex.

We can also define complex variables with the instruction complex(a,b) where a and b are the real and imaginary parts, respectively. The complex conjugate of a complex number $z = x + iy$ is given by

$$z^* = x - iy$$

To obtain the complex conjugate of a complex number z we use the instruction conj(z)

```
>> conj(z) % returns the complex conjugate of z

ans =
    x-i*y

>> conj(x)

ans =
    x
```

Since x is real, conj(x) returns the same value. The product of a complex number z by its complex conjugate is its magnitude squared

```
>> mag_sq = z*conj(z)

mag_sq =
(x+i*y)*(x-i*y)
```

With the instruction expand, MATLAB rewrites the results as

```
>> magSq = expand(mag_sq)

magSq =
x^2+y^2
```

And in a mathematical format with

>> **pretty(magSq)**

$$x^2 + y^2$$

A variable that has been defined as complex can be made real with

syms x real

2.2   Functions

There exist two types of functions in MATLAB: those already predefined in MATLAB and those defined by the user. Some of MATLAB functions fall within the set of elementary functions such as trigonometric, logarithmic, and hyperbolic functions among others. Some elementary functions were presented in Chapter 1.

As mentioned above, it is possible to have user-defined functions. We can define them in the Command Window or in a script in the MATLAB editor. To write a script it is necessary to open the MATLAB editor and click on the New M-File icon. We write the instructions in the window open for the new file. When finished, we save the file, name it with an appropriate name, and run it with the Run icon or with the F5 key, or from the Command Window entering the file name. All files corresponding to scripts and functions are saved with the extension .m. We now show two examples, in the first one we enter all the instructions in the Command Window and for the second one we write a script.

Example 2.1 Calculation of the areas and perimeters of a hexagon and a triangle

We wish to calculate the area and perimeter for a hexagon and a triangle. The triangle is equilateral with side 8. The hexagon side is 10. The equation to calculate the area of a triangle with base L is

$$A_T = \frac{Lh}{2} \tag{2.3}$$

where h is the triangle's height. The height h is given by $h = L\sqrt{3}/2$. For the hexagon, the area is given by

$$A_H = \frac{3\sqrt{3}}{2}L^2 \tag{2.4}$$

and the perimeter is

$$P_H = 6L \tag{2.5}$$

where L is the side of the hexagon. In the Command Window we enter the following instructions:

```
>> Lt = 8; Lh = 10; %Triangle side = 8, hexagon side = 10
>> h = sqrt(3)/2; % triangle's height.
>>% Now we enter the equations for the calculations needed.
>> Area_triangle = Lt*h/2
    Area_triangle =
        3.4641
>> Area_hexagon = 3*sqrt(3)*Lh/2
Area_hexagon =
        25.9808

>> Perimeter_hexagon = 6*Lh

Perimeter_hexagon =
        60.0000
```

In the previous example we have placed comments after the instructions. Comments can thus start at the end of the instruction or at the beginning of the row.

Example 2.2 Ball throwing

A player throws a ball. The ball trajectory makes an angle θ with respect to the horizontal. The initial velocity is v. We want to find the total distance traveled by the ball before it falls to the ground and the time the ball is in the air. The velocity vector components are

$$v_{ox} = |v| \cos \theta \tag{2.6}$$

$$v_{oy} = |v| \sin \theta \tag{2.7}$$

It can be shown that the time t the ball is in the air, the distance d travelled, and the ball path are given by

$$t = \frac{2v_{oy}}{g} \tag{2.8}$$

$$d = v_{ox}t \tag{2.9}$$

$$y = \frac{v_{oy}}{v_{ox}}x - \frac{1}{2}\frac{g}{v_{ox}}x^2 \tag{2.10}$$

Here g is the acceleration due to gravity. The data is $\theta = 40°$, $|v| = 45$ m/s. We can write a script to perform the desired calculations. The first thing to

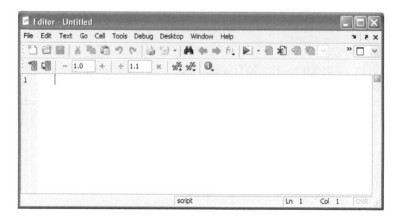

Figure 2.1: MATLAB Editor window.

do is to click on the New M-File icon in the left top corner of the MATLAB window. This opens the MATLAB Editor with an empty window to write our data, equations, and comments. The Editor window is shown in Figure 2.1.

The instructions to do the calculations are written in the following m-file:

```
theta = 40; % initial angle in degrees.
theta_rad = theta*pi/180; % degree to radians conversion.
velocity = 45; % initial speed.
g = 9.8; % acceleration due to gravity.
v0x = velocity*cos(theta_rad);
v0y = velocity*sin(theta_rad);
time = 2*v0y/g;
distance = v0x*time;
fprintf('Time elapsed %12.5f \n', time)
fprintf('Distance travelled by the ball %12.5f \n ', distance)
x = 0: 0.1: distance;
y = (v0y/v0x)*x-(1/2)*(g/v0x^2)*x.^2;
plot(x, y)
xlabel('Distance')
ylabel('Height')
grid on
```

We save the script as file ball_path.m and we run it to obtain the following results:

```
Time elapsed 5.90315
Distance travelled by the ball 203.49344
```

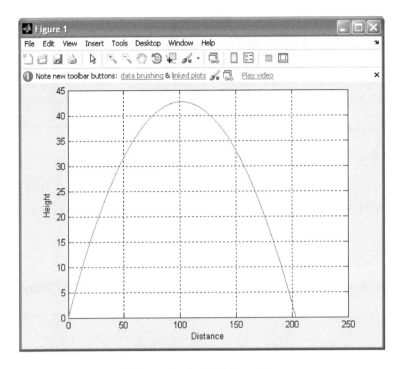

Figure 2.2: Ball's path.

The path's plot is shown in Figure 2.2.

2.2.1 MATLAB Elementary Functions

The elementary functions are grouped in the set Elementary Functions, elfun. The list can be displayed with

>> **help elfun**

In this way we obtain a list of all the elementary functions in this set. There we can find all of the trigonometric, hyperbolic, logarithmic, exponential, complex, and others for rounding, residue, and sign functions.

Each one of these functions can be used with only typing the name with the argument between parentheses. For example, for the square root of 2 and the sine of π we use

sqrt(2)
sin(pi)

2.2.2   Using Symbolic Functions

There are some operations that we can do on symbolic functions. Thus, we can apply the elementary functions to functions using numeric or symbolic variables. For example, let us suppose that we want to find the first derivative of the function f defined by

$$f(x) = x^3 + 2x \tag{2.11}$$

To define the function in a symbolic manner we can use

syms x
f = sym('x^3+2*x')

> f =
> x^3+2*x

To evaluate a symbolic function in a given variable value we use the instruction

subs(f, a, value_of_a)

For example, for the function defined above, if we want to find its value when x = 2 we use

>> **Value_of_f = subs(f, x, 2)**

Value_of_f =
    12

Now if we want to find the derivative we use the instruction diff(f, n) where n indicates what order derivative we want to find. For the function f we have then,

derivative_f = diff(f)
derivative_f =
    3*x^2+2

and to write it in a mathematical form we use the instruction pretty as

pretty(derivative_f)
       2
    3 x + 2

To integrate we use the instruction int(f). Thus for f we have,

integral_f = int(f)

    integral_f =
    1/4*x^4+x^2

pretty(integral_f)
      4   2
    1/4 x + x

And the definite integral from 1 to 2 can be found with

int(f, 1, 2)
    ans =
    27/4

2.2.3 Plots

The instruction **plot** is useful to plot numeric vectors as we saw in Chapter 1. In the case of symbolic functions, the instruction we have to use is **ezplot**. For example, to plot the function $f(x) = x^2 \sin(x)$, we use

    f = sym('x^2*sin(x)')
    ezplot(f)

And we obtain the plot shown in Figure 2.3. We can also add limits to this instruction. If we want to plot from x = -10 to x = 10 we use

    ezplot(f, -10, 10)

The resulting plot is shown in Figure 2.4.

    Another instruction we can use is **fplot**. The same format applies to both **ezplot** and **fplot**. Thus, for the plot of the function

$$e^{-x/2} \sin^2(x)$$

in the range from -2 to +4 we can use (in **fplot** we have to add the range). Thus, we have

    fplot('exp(-x/2)*sin(x)^2', [-2, 4])

This produces the plot shown in Figure 2.5.

    Note the difference among the functions **plot**, **ezplot**, and **fplot**. The function **plot** must have the data in vector form while **ezplot** and **fplot** work on a symbolic function. That is, we have to generate the numeric data for **plot** in the form of two vectors, one vector for the independent variable and another one for the dependent variable.

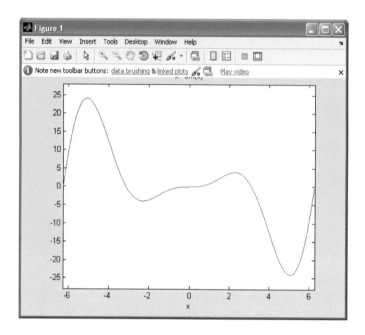

Figure 2.3: Plot of the symbolic function $f(x) = x^2 * \sin(x)$.

2.2.4   Function Evaluation Using **eval** and **feval**

To evaluate a symbolic variable or function we can use the instruction eval. For example, to evaluate the function sin(x) at $x = \pi/2$, that we know it has unity value, we can use

```
eval ('sin(pi/2)')
ans =
    1
```

The symbolic function must be between single quotes in the same way strings are treated besides being a valid MATLAB expression.

Another useful evaluation instruction is feval. This is used to evaluate symbolic expressions at a given number of points. For example, let us suppose that we wish to evaluate the function tan(x) at the points $x = 1$, 2, 3. We do this by first defining x as $x = [1, 2, 3]$ and then we use feval('tan', x). Thus,

```
x = [1, 2, 3]
feval('tan', x)
    ans =
    1.5574 -2.1850 -0.1425
```

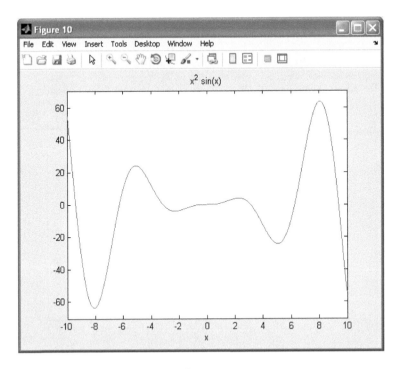

Figure 2.4: Plot of $f(x) = x^2 * \sin(x)$ in the range -10 to 10.

2.2.5 The Function funtool

MATLAB has a tool to work with functions and observe their plots. It can work with two functions called $f(x)$ and $g(x)$. This tool has the name funtool (function tool) and it is started by writing in the Command Window

    funtool

Three windows belong to funtool. They are numbered as parts a, b and c and are shown in Figure 2.6. In Figure 2.6c we can enter the functions f(x) and g(x) as well as a constant a and a set of values for x. This window has a keyboard to perform calculations on the functions. We can use the keyboard to find the derivative of the functions, the composite function f[g(x)] the product of them, etc. When starting funtool, Figure 2.6c has by default the functions f(x) = x, g(x) = 1, but these functions can be changed by users. Also the constant value is a = 1/2 and the range of x is [-2 , 2] . As an example, if

$$f(x) = \cos^3 x/(1 + x^2), \; g(x) = 1/(x + 1)^2, \; a = 2, \; x = [-2\pi, 2\pi]$$

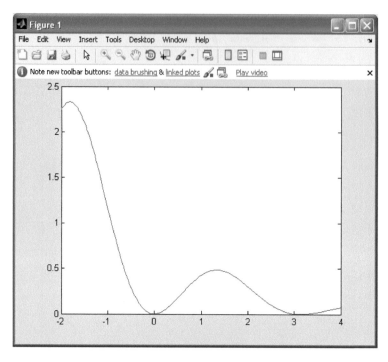

Figure 2.5: Plot of $f(x) = \exp(-x/2) * \sin^2(x)$ in the interval [-2, 4].

we get the plots shown in Figure 2.7.

2.3 Polynomials

Polynomials are a very special class of functions. A polynomial is a function of the form

$$p(x) = a_n x^n + a_{n-1} x^{n-1} + \ldots + a_1 x + a_0 \tag{2.12}$$

The quantities a_k are called the polynomial coefficients. In MATLAB a polynomial is represented as row vectors and the coefficients are given in descending order. For example, for the following polynomials we have

$$
\begin{array}{lll}
x + 1 & \text{is} & [1 \ 1] \\
x - 1 & \text{is} & [1 \ \text{-}1] \\
4x^2 + 2x - 3 & \text{is} & [4 \ 2 \ \text{-}3] \\
-2x^2 + \sqrt{(7)}x + 5 & \text{is} & [\text{-}2 \ \sqrt{(7)} \ 5]
\end{array}
$$

For the last example, we have

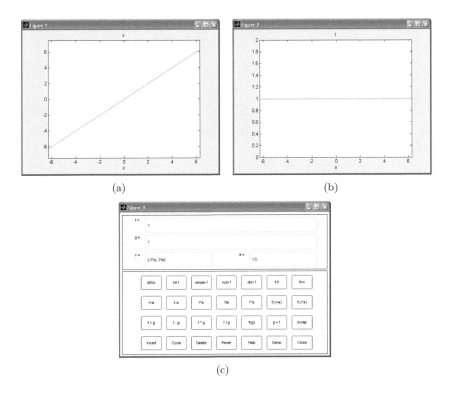

Figure 2.6: **funtool** windows. (a) **Window for** $f(x)$, (b) **Window for**
$g(x)$, (c) **Main window. Here the user specifies the functions and**
operations.

poly = [-2 sqrt(7) 5]
poly =
-2.0000 2.6458 5.0000

The instruction **fliplr**, which means flip from left to right, changes the order
in which a polynomial is written. Thus, **poly_1** = $[1\ 2\ 3\ 4]$ is the polynomial
$x^3 + 2x^2 + 3x + 4$, then

poly_2 = fliplr(poly_1)
poly_2 =
    4 3 2 1

represents the polynomial $4x^3 + 3x^2 + 2x + 1$.

To evaluate a polynomial at a given value of x we can use the instruction
polyval(poly, x). For the polynomial given by **poly** above, its value at $x = 9$ is

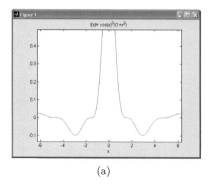 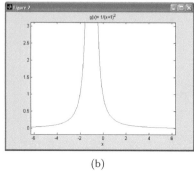

(a)   (b)

Figure 2.7: funtool **windows. (a) Window for** $f(x) = \cos^3 x/(1 + x^2)$**,**
(b) Window for $g(x) = 1/(x + 1)^2$**.**

polyval(poly, 9)
ans =
    -133.1882

We can change the single valued x for a vector of x values as in

x = [1 -1 2 5 6 9];
polyval(poly, x)

ans =
    5.6458 0.3542 2.2915 -31.7712 -51.1255 -133.1882

To plot a polynomial we simply define the vectors x, y as in

x = linspace(0, 2, 100);%Definition of the vector x.
poly = [6 3 -7 0.4];% Definition of the polynomial.
y = polyval(poly, x);% Definition of the vector y.
plot (x, y)% Plotting of y vs. x.
grid on

The resulting plot is shown in Figure 2.8.
    To find the roots of a polynomial we have the instruction **roots**. For ex-
ample,

zeros = roots(poly)
zeros =
    -1.3803
    0.8215
    0.0588

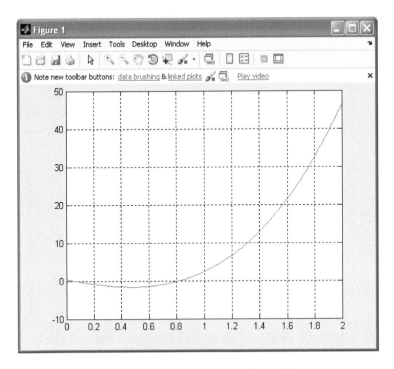

Figure 2.8: Plot of polynomial $6x^3 + 3x^2 - 7x + 0.4$.

Since the polynomial is of third degree, we obtain three roots, also called zeros. Note that the roots are in a column vector while the coefficients are in a row vector. We can plot the zeros with

stem (zeros)
title ('Stem plot for the zeros of the polynomial')

The stem plot of Figure 2.9 is a special type of plot for data points as used in discrete mathematics. We have changed the x limits in the plot with the menu Edit→ Axis Properties and changing there the limits from 0.0 to 4.0.

If we have a set of roots, we can obtain the original polynomial with the instruction poly. For the polynomial given we have

a = poly(zeros)
a =
      1.0000 0.5000 -1.1667 0.0667

Note that the leading coefficient is unity. Multiplying this result by 6 we get the original polynomial as

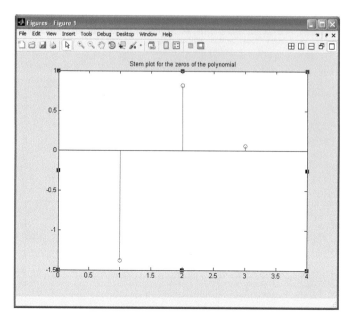

Figure 2.9: Stem plot for the roots of polynomial $6x^3 + 3x^2 - 7x + 0.4$.

a = 6*a
a =
    6.0000 3.0000 -7.0000 0.4000

Note that both polynomials have the same roots. It may be possible in some cases to have a small imaginary part. This is due to round off errors and it should be neglected. This happens because MATLAB works with complex numbers. We can use the instruction real to eliminate any imaginary part.

The product of two polynomials is another polynomial. We can use the instruction conv to multiply two polynomials as in

conv (polynomial_1, polynomial_2)

For example:

poly_2 = [3 -4 7];
poly_1 = [4 -2 0 1];
conv(poly_1, poly_2)

ans =
    12 -22 36 -11 -4 7

which corresponds to the polynomial

$$p(x) = 12x^5 - 22x^4 + 36x^3 - 11^2 - 4x + 7 \qquad (2.13)$$

With conv we can only multiply two polynomials at a time. If we need to multiply three polynomials we first multiply two and then the result is multiplied with the remaining one. If we want to multiply $(x^2-x-1)(x^3-2)(x^2-3)$ we can use

```
poly_1 = [1 -1 -1];
poly_2 = [1 0 0 -2];
poly_3 = [1 0 -3];
pl1 = conv(poly_1, poly_2);
poly_final = conv(pl1, poly_3);
    poly_final =
    1 -1 -4 1 5 8 -6 -6
```

This can also be calculated with

```
pm = conv(poly_3, conv(poly_1, poly_2))
    pm =
    1 -1 -4 1 5 8 -6 -6
```

Both results give the polynomial

$$p_m(x) = x^7 - x^6 - 4x^5 + x^4 + 5x^3 + 8x^2 - 6x - 6$$

To divide polynomials we use the instruction deconv that has the format

$$[q, r] = \text{deconv }(a,b)$$

The polynomials q and r are the quotient and the residue of the division, respectively. For example, for the polynomials a and b we have

```
a = [1 6 24 50 77 84 64];
b = [1 4 9 16];
[q, r] = deconv(a, b)
    q =
    1 2 7 -12
    r =
    0 0 0 0 30 80 256
```

That is, the quotient and residue polynomials are

$$q(x) = x^3 + 2x^2 + 7x - 12$$

and

$$r(x) = 30x^2 + 80x + 256$$

To add polynomials they must have the same degree. If they do not have the same degree, then we have to add zeros to the lower degree polynomial. For the polynomials a and b given above we have to add zeros to polynomial b as in

a = [1 6 24 50 77 84 64];
b = [0 0 0 1 4 9 16];
c = a+b
   c =
   1 6 24 51 81 93 80

Then the polynomial $c(x)$ is

$$c(x) = x^6 + 6x^5 + 24x^4 + 51x^3 + 81x^2 + 93x + 80$$

The derivative of a polynomial can be found with the instruction

polyder (polynomial)

For polynomial a we have

polyder(a)
   ans =
   6 30 96 150 154 84

A rational function is the quotient of two polynomials. That is, it is of the form

$$\frac{polynomial\ \ p(x)}{polynomial\ \ q(x)}$$

A rational function can be expanded in partial fractions. This is an expansion of the rational function to an expression of the form

$$\frac{p(x)}{q(x)} = \sum \frac{k_i}{x + p_i} + k_\infty x \tag{2.14}$$

In this expansion, p_i is called the pole and k_i is the residue at pole p_i and k_∞ is the residue due to a pole at infinity. For example, for the polynomials $p(x)$ and $q(x)$ given by

$$p(x) = 10x + 20$$

$$q(x) = x^3 + 8x^2 + 19x + 12$$

The partial fraction expansion is given by

$$\frac{10(x+2)}{x^3 + 8x^2 + 19x + 12} = \frac{10(x+2)}{(x+4)(x+3)(x+1)} = \frac{-6.67}{(x+4)} + \frac{5}{(x+3)} + \frac{1.667}{(x+1)}$$

The poles of the rational function are $p_i = $ -1, -3, -4, the residues at these poles are $k_i = $ -6.667, 5, 1.667. Since the rational function does not have neither a pole at infinity nor a pole at the origin, $k_\infty = 0$.

In MATLAB we can calculate the poles and residues with the instruction

[residues, poles, k_inf] = residue (poly_numerator, poly_denominator)

The results of this instruction are a vector with the residues, a vector of poles, and the residue for the pole at infinity. For example,

num = [10 20];
den = [1 8 19 12];
[residues, poles, k] = residue(num, den)
    residues =
    -6.6667
    5.0000
    1.6667
    poles =
    -4.0000
    -3.0000
    -1.0000
    k =
    []

The same instruction is used for the inverse operation. That is, if we give the poles, residues, and constant term we can obtain the polynomials of the rational function with the instruction residue as

[numerator, denominator] = residue (residues, poles, k)

For example, for the data given below we obtain

residues = [1 2 3];
poles = [-1+j, -1-j, -2];
k_inf = 2;
[n, d] = residue(residues, poles, k_inf)
    n =
    2.0000 14.0000 27.0000 - 1.0000i 20.0000 - 2.0000i
    d =
    1 4 6 4

The numerator and denominator polynomials are then

$$n(x) = 2x^3 + 14x^2 + (27 - i)x + (20 - 2i)$$

$$d(x) = x^3 + 4x^2 + 6x + 4$$

2.4   Curve Fitting

In many scientific and engineering applications it is necessary to describe experimental data in an analytical form by means of a function. Thus, we can describe the experiment by a function which in many cases it is a polynomial. MATLAB allows us to fit data to a polynomial by using the instruction polyfit which has the format

p = polyfit(x, y, n)

where x,y are data vectors and n is the order of the polynomial p. For example, let us assume that we collected the data in Table 2.1 from an experiment. We can plot this data, as follows:

Table 2.1: Experimental data

x	y
0	0
1	1
2	3.3
3	2.2
4	5.6
5	4.4
6	0

x = [0, 1, 2, 3, 4, 5, 6];
y = [0, 1, 3.3, 2.2, 5.6, 4.4, 0];
plot (x, y, '*k')

The plot is shown in Figure 2.10. We have changed the x axis limits from -1 to 7 using from the main menu Edit→ Axis Properties. The symbol '*k' indicates that the points are plotted with asterisks. We now use the instruction polyfit to find a third-degree polynomial that approximates the data points with:

p = polyfit(x, y, 3)
p =
   -0.1583 1.0024 -0.3060 0.2190

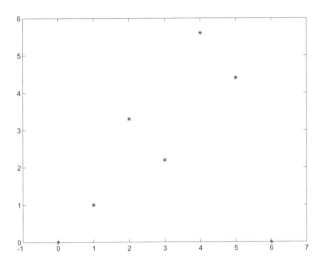

Figure 2.10: Plot of experimental data from Table 2.1.

That is, the polynomial we get is

$$p(x) = -0.1583x^3 + 1.0024x - 0.3060x + 0.2190$$

We can plot the polynomial $p(x)$ together with the data points with

```
x1 = linspace(0, 7, 100);
p1 = polyval( p, x1);
plot(x, y,'*r', x1, p1)
```

This produces the plot of Figure 2.11. We see that the polynomial is a good approximation to the data points. We now try a fifth-order polynomial to see that the higher the order the better the approximation. Then,

```
p5 = polyfit(x, y, 5)
  p5 =
  -0.0087 0.0718 -0.1301 -0.1960 1.6957 -0.0618
```

and we now plot it together with the data points and the third-degree polynomial with

```
p5v = polyval(p5, x1);
plot(x, y, '*r', x1, p1, x1, p5v, '-.')
legend('Data points', '3rd order', '5th order')
```

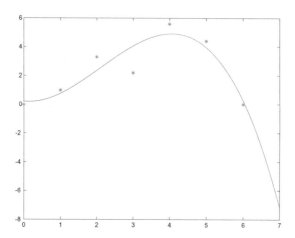

Figure 2.11: Plot of experimental data from Table 2.1 and third-degree interpolation polynomial.

A plot showing data points, the third-order polynomial, and the fifth-order polynomial are shown in Figure 2.12. We can readily see that the fifth-order polynomial is a better approximation to the data points.

2.4.1   Cubic Spline Fitting

A cubic spline provides another way to curve fitting. A cubic spline is a third-degree polynomial that exactly passes through the data points and in addition its first derivative is continuous at the data points. The instruction to obtain a cubic spline is

$$z = \textsf{spline (data\_points\_x, data\_points\_y, x)}$$

where data_points_x and data_points_y are the coordinates of the data points, and x is a vector of points where we wish the spline z to be evaluated. For example, for data points input_x, input_y we want to find the spline, in addition we wish the spline evaluated at the points given by the vector x defined below

```
input_x = [1 2 3 4 5 6];
input_y = [1 0 4.4 0 5.5 0];
x = linspace(1, 6, 100);
z = spline(input_x, input_y, x);
plot(input_x, input_y,'*r', x, z)
```

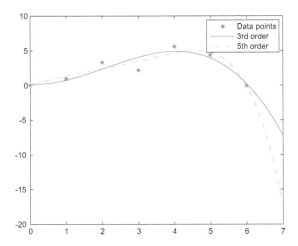

Figure 2.12: Plot of data points and third- and fifth-order polynomials.

The variable z contains the spline data evaluated at the points in vector x. The plot of the data points and the spline is shown in Figure 2.13. We readily see that the spline provides an exact approximation at the data points. This is due to the fact that the spline is in fact a set of third-degree polynomials.

2.4.2 The Tool **Basic Fitting**

We can also do a curve fitting using the MATLAB tool called **Basic Fitting**. This tool is available with the *Curve Fitting Toolbox*. It can be called from any plot available. For example, if we have sets of data points x, y, we just plot them. In the figures Tools menu we select the option for **Basic Fitting** as shown in Figure 2.14. This selection opens the window of Figure 2.15 which gives the option to select the type of approximation we wish to do. We can choose from the different degrees of polynomial approximation or spline approximation. As mentioned above, the best type of approximation is a cubic spline, but this option does not give an analytical expression as does the polynomial approximation. Figure 2.16 gives the results of a spline and a fifth-order polynomial approximation. If we plot the residuals, we get the error at the bottom of Figure 2.16 as shown there. The residuals are the error between the polynomial value at the x, y points and the actual polynomial value at the x, y points. A closer look at Figure 2.15 shows an arrow at the bottom right corner. This arrow is to open a second and third section in this figure as shown in Figure 2.17. The second section of the window shows the polynomial coefficients and the norm of the residuals. The third window is

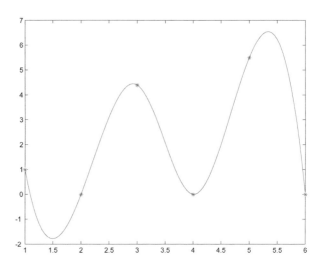

Figure 2.13: Plot of data points and the spline.

only to evaluate the polynomial at a given value of the variable x. Finally, with a click on the left pointing arrow we can close the sections.

2.5   Solution of Equations

To solve equations we can use the instruction solve. The argument for this instruction can be an equation such as

$$ax^3 + bx^2 + cx + d = 0 \qquad (2.15)$$

Or simply an expression as

$$\sin(x) + 2 \qquad (2.16)$$

In this case, MATLAB equates to zero the expression. As examples we have,

```
solve('a*x^2+b = 0')
    ans =
    1/a*(-a*b)^(1/2)
    -1/a*(-a*b)^(1/2)

solve('a*x+b', 'x')
    ans =
    -b/a
```

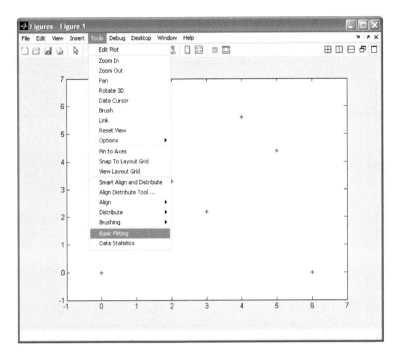

Figure 2.14: Selection of the tool **Basic Fitting**.

Figure 2.15: Main window for the tool **Basic Fitting**.

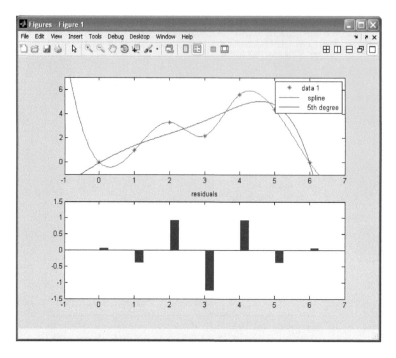

Figure 2.16: **Curves for the fifth-degree polynomial and spline and plot for error approximation (residuals).**

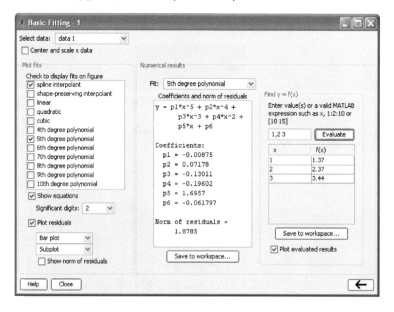

Figure 2.17: **Basic Fitting** window with coefficients window and evaluation of the function.

Note that in the second example we have indicated with 'x' that the unknown is x. We also see that even though in the first case we did not indicate what is the variable, MATLAB takes as unknowns the last few letters of the alphabet. Thus, the unknown is taken as x. If the unknown is either a or b we have to indicate this explicitly. If for example the unknown is a, we indicate this as

```
solve('a*x+b', 'a')
    ans =
    -b/x
```

In some cases we get a numeric result as in

```
f = solve('cos(x) = sin(x)')
    f =
    1/4*pi
```

We get a numeric value with the instructions numeric and eval:

```
numeric(f)
    ans =
     0.7854
eval(f)
    ans =
    0.7854
```

If MATLAB cannot find a symbolic solution, then it gives a solution that looks like a number as in

```
x = solve( 'exp(x) = tan(x)' )
    x =
    1.3063269404230792361743566584407
```

This is not a numeric value, rather it is a symbolic one. To evaluate this symbolic result we use eval as in

```
eval(x)
    ans =
    1.3063
```

We can also solve equations with the instruction fzero. This function uses a function defined in an m-file. The format for the instruction fzero is

$$x\_sol = fzero(function, x\_ini)$$

where x_sol is a vector with all the possible solutions of the equation

$$\text{function} = 0$$

x_ini is an initial value for the solution. For example, to find the solutions of

$$(x - \pi)x = 7$$

We have that the equation can be written as

$$f(x) = (x - \pi)x - 7$$

Using the MATLAB editor we write the following m-file and save it,

```
function y = fx(x);
% This is file fx.m
y = (x-pi).*x-7;
```

Giving the initial value as x = 4, we now run the instruction

```
fzero('fx', 4)
ans =
4.54138126514911
```

If we change the initial condition we get another solution as

```
fzero('fx', -1)
ans =
-1.54138126514911
```

A way to find initial values is to plot the function and from there make an educated guess about the initial values by observing the zero crossings. We can plot the functions with the instructions fplot and ezplot as in

```
ezplot( 'function')
fplot('function')
```

For example, to find the roots of

$$(x - 2)^3 - \sin(x) = 1$$

We use

```
syms x
ezplot((x-2)^3-sin(x))-1)
grid on
```

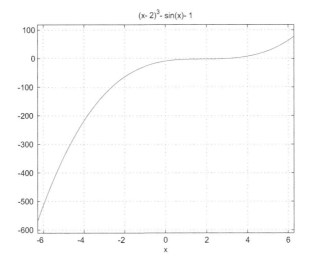

Figure 2.18: Plot of the function $f(x) = (x-2)^3 - \sin(x) - 1$.

This gives the plot of Figure 2.18. We see that there is a zero crossing close to x = 2. We use this as an initial condition. Then we use

root = fzero('(x-2)^3-sin(x)-1', 2)
  root =
  3.0344

We see that the solution is x = 3.0344.

2.6 Execution Time, Date, and Time of the Day

There are several functions that allow us to find the execution time, the date, and the Time of the Day. In this section we cover the instructions cputime, clock, tic, toc, date, datenum and now.

The instruction cputime indicates for how long MATLAB has been working in the current session. The response is given in seconds. If we execute this instruction we get

cputime
  ans =
  28.1738

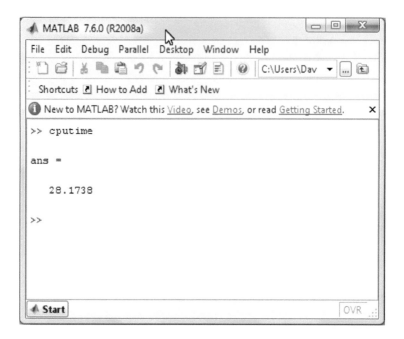

Figure 2.19: Seconds from the start of the current MATLAB session.

which indicates that the current session started 28.1738 seconds ago. Figure 2.19 shows the time elapsed from the beginning of the session. We can use cputime to find the time taken to execute a set of instructions. For example,

```
t_initial = cputime;
x = linspace (0, 0.1, 100);
a = sin (x);
t_final = cputime - t_initial
    t_final =
    0.2500
```

Thus, we see that the time elapsed in the above calculations was 0.25 seconds. This time will change for different computers.

The instructions tic and toc are used to measure the time elapsed in the execution of a set of instructions. For example,

```
tic;
x = linspace (0, 0.01, 100);
x = tan (x).* sin (x);
toc;
```

Elapsed time is 0.026694 seconds.

The time elapsed depends upon the processor speed and on the number of programs running at the same time.

With the instruction clock we obtain a row vector with the date and time. The data format is: year, month, day, hour, minutes, second.

clock
```
ans =
1.0 e+003 *
    2.0090 0.080 0.0200 0.0200 0.0010 0.0224
```

which indicates that the year is 2009, the eighth month August, the 20th day, and the time is 20 hr or 8 PM, 1 minute, and 22.4 seconds.

The instruction date gives the date in a string,

date
```
ans =
20-Aug-2009
```

As with any other string, we can write only a few characters of the string

string = date;
string (3:7)
```
ans =
-Aug-
```

MATLAB counts the days beginning in January 1st, 0000. This is only a reference point. The instruction datenum displays the code number for any date in particular. For example, the date January 14, 2010 gives:

n = datenum(2010, 01, 14)
```
n =
734152
```

If we wish to know how many days have passed between two dates we can use datenum. For example, to know how many days have passed between October 5, 2011 and October 5, 2012 we use

n = datenum('10-5-2012') - datenum('10-5-2011')
```
n =
366
```

And we see that 2012 is a leap year.

The format for the date can be one of several ones. For example

datenum('January-14-2010')
datenum('january-14-2010')
datenum('14-jan-2010')
datenum('1/14/10')
datenum('14-jan-2010')
    ans =
    734152.00

All of them give the same result.

The hour can also be included with **datenum** as is

format bank
datenum('january-14-2010 12:00 PM')
    ans =
    734152.50

The decimal part is due to the fact that 12 PM is a half of the day.

The instruction **now** gives the date in a numeric format. For the day May 10, 2010 we get

now
    ans =
    734268.45

2.7   Concluding Remarks

We have presented the way MATLAB works with variables and functions, either user-defined or MATLAB predefined functions. Even though MATLAB has more that ten thousand predefined functions, in most cases it will not have a particular function needed by a user. Thus, it is of paramount importance that users know how to define their own functions. We also presented how we can plot symbolic functions. The form to represent polynomials in MATLAB is important because they can be represented as vectors. Data fitting is presented and several examples show how we can interpolate data to a polynomial or to a spline. The tool Basic Fitting provides an easy-to-use interface to data fitting. Finally, we presented the way to measure execution time and data and time formats.

2.8 Exercises

Section 2.1

2.1 Define a variable called university. Give it the value 5.10+i*1989. Get its real and imaginary parts.
2.2 Write the variable a=[1 9 5 0] in the Command Window and see it in the Workspace Window. Select it in the Workspace and click on the Plot icon. This action must plot the set of points joined by straight lines.
2.3 Extract the real and imaginary parts for the function

$$f(x) = e^{3i/4}$$

Section 2.2

2.4 Write a script to find the volume of a cone whose base radius is 3 and its height is 27. The volume of a cone is given by

$$V = \frac{\pi r^2 h}{3}$$

2.5 Write a script to perform synthetic division of two polynomials. The method can be found on any algebra book. Then divide the polynomial $p(x) = 3x^5 + 4x^3 - 5x^2 - 7x + 9$ by the polynomial $q(x) = 6x^2 - 8x + 2$.
2.6 Write a script to find out the solution of a second order equation. Then find the solution for

$$x^2 + 0.1x + 2 = 0$$

2.7 Obtain the derivative for the function

$$f(x) = x - 5x^{cos x}$$

2.8 Evaluate the integral

$$\int t^2 e^{-t}$$

2.9 Evaluate the integral of the function $1/x$ between the limits 2 and 20.
2.10 Find the derivative of the function $a/(x - b)$ with respect to x.

Section 2.4

2.15 Write in MATLAB the polynomial

$$p(x) = x^3 + 2x^2 + 2x + 1$$

and find its value at $x = 2$.
2.16 Plot the polynomial

$$p(x) = 16x^5 - 20x^3 + 5x$$

in the rank from $x = $ -1.5 to +1.5.

2.17 Find and plot the roots for the polynomial

$$d(x) = x^5 + 3.236x^4 + 5.236x^3 + 5.236x^2 + 3.236x + 1$$

2.18 Multiply the polynomials

$$p(x) = 2x^5 + x^4 + 3x^3 + 2x^2 + 6x + 1$$

$$q(x) = -10x^3 + 2x^2 - 7x + 8$$

2.19 Evaluate the quotient of the polynomials

$$p(x) = 2x^6 - 7x^5 + 6x^4 - 9x^3 + 4x + 8$$

and

$$q(x) = x - 8$$

2.20 Find the derivative for the polynomial

$$p(x) = 23x^6 + 17x^5 - 4x^4 + 67x^3 - 14x + 1$$

2.21 Form the ratio of the polynomials and find the poles and residues using the instruction residue for

$$p(x) = 2x^5 + 6x^4 + 3x^3 + 3x^2 + 6x + 2$$

and

$$q(x) = -7x + 4$$

2.22 For the data in Table 2.1 obtain an approximating second-degree polynomial that fits the data.

2.23 Repeat Exercise 2.22 but for: (a) a fourth-degree polynomial, (b) a sixth-order polynomial. Which answer gives a better approximation?

2.24 Obtain a spline approximation for the data of Exercise 2.22.

Section 2.5

2.25 Solve the equation

$$-17x^5 + 66x^4 - 89x^3 - 14x^2 + 81 = 0$$

2.26 Solve the equation

$$7x^3 - 7x^2 - 8 = 0$$

2.27 Solve the equation $77x^3 - 4 = \sin(x)$ using the instruction fzero.

Section 2.6

2.28 Find the execution time for the execution of the functions: (a) $\sin(\pi/3)$, (b) $\sqrt{2i}$, c) $e^{(7i/12)}$.

2.29 Obtain the current time and date.

2.30 Obtain the current date in MATLAB format.

2.31 Write the current date and time using the instruction now.

2.9 References

[1] R. Larson, R. P. Hostetler, B. H. Edwards, Precalculus Functions and Graphs: A Graphing Approach, 5th Ed., Brooks/Cole Thomson, Pacific Grove, CA, 2007.

[2] J. W. Brown and R. V. Churchill, Complex Variables and Applications, McGraw-Hill Book Co., New York, 2008.

[3] B. Kolman and D. Hill, Elementary Linear Algebra with Applications, 9th Ed., Prentice Hall, Inc., Piscataway, NJ, 2008.

[4] Symbolic Math Toolbox Users Guide, The MathWorks, Inc., Natick, MA, 2009.

[5] I.M. Gelfand and A. Shen, Algebra, Birkhäuser, Boston, MA, 2003.

Chapter 3

Matrices and Linear Algebra

As mentioned in Chapter 1, MATLAB was originally designed to carry out operations with matrices, and received the name of Matrix Laboratory that subsequently was reduced to MATLAB. Thus, it deserves a special attention to learn the way in which MATLAB works with them.

Matrices are practically found in all areas of knowledge. In this manner, it is possible to find applications of matrices not only in the mathematics areas, but also they find a great deal of application in engineering, in physics, in finances, in accounting, in arts, in music, in anthropology, chemistry and biology, to mention a few.

From the definition of a matrix given previously, we can see why a matrix is also called an array.

MATLAB has an array editor that allows users to introduce the elements of a matrix or to modify the elements of a matrix given previously and already stored as a MATLAB variable. This array editor will be described later in the chapter.

Another important application of matrices is given in the solution of systems of simultaneous equations because the equations' coefficients are arranged in a matrix. Simultaneous equations systems appear in many problems in engineering such as in optimization of systems, automatic control, food engineering, chemical engineering, etc. Chapters 9, 10 and 11 present several matrix applications in the solution of problems.

A vector is a matrix with either only a column or a row. Thus, a vector is a particular case of a matrix.

In addition to the basic operations with matrices and vectors, MATLAB has the potential to handle arrays by means of indexing. In this manner carrying out calculations using arrays becomes very efficient and fast.

Finally, the chapter covers cell and structure arrays. Cell arrays are also matrices, but the fundamental difference is that they can handle a combination of different variable types such as names of people or things and numbers. These arrays allow us to handle data in a more efficient and more practical way. A structure is similar to a cell but each element is identified by a name.

The chapter begins with a description of matrices and the basic operations that can be performed with them. It continues with vectors and the way to compute dot and vector product. A section is dedicated to matrix and vector functions that can be carried out on matrices and vectors. A section is presented to solve simultaneous equations systems, and for LU factorization of a matrix. The chapter ends with cell arrays and structures.

3.1   Matrices

MATLAB handles in matrix form all the variables defined in a MATLAB session, whether they are user-defined or they are in the predefined functions inside MATLAB. It is then convenient to consider the concept of a matrix that any book of linear algebra defines as:

A Matrix is an array of numbers or objects.

In this way, a matrix can be an array of numbers, letters, objects, or any other things. The only condition that a matrix should satisfy to be a matrix is for it to be an ordered array. For example, a table of numbers is a matrix. The pixels of an image also form a matrix. These two matrices are ordered in a plane. If an image changes with time, as is the case of television images or motion pictures, then we are adding another dimension to the definition of such matrices and thus we are talking about multidimensional matrices. The objects that form the matrix are called **elements** of the matrix.

An example of a matrix is

$$A = \begin{bmatrix} 1 & -3+2i & 0 \\ 8 & -6 & 4 \\ 2 & 27 & 5 \\ -54 & 10 & 25 \end{bmatrix}$$

which has 4 rows and 3 columns. We say that matrix A is of dimension 4×3. If a matrix has n rows and m columns, it is a $n \times m$ matrix. The quantity $n \times m$ is called the dimension of the matrix. In MATLAB, the matrix A is defined by

$$A = [1\ \text{-3+2i}\ 0;\ 8\ \text{-6}\ 4;\ 2\ 27\ 5;\ \text{-54}\ 10\ 25]$$

That is, we have separated the elements of a same line by a space (they can be also separated with a comma) and the rows are separated with a semicolon. Each element from the matrix A has a position i, j which corresponds to its position in the ith row and in the jth column. For example, in matrix A the

Table 3.1: Matrix Commands

$A(a{:}b,c{:}d)$	Submatrix consisting of rows a to b and by columns c to d.
$A(k,:)$	Vector formed by the elements in the k-th row.
$A(:,j)$	Vector formed by the elements in the j-th column.

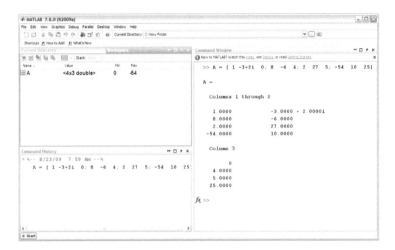

Figure 3.1: Matrix A in the Command Window.

element 5 has the position 3,3 while the element 27 has the position 3,2. The command

A(i, j)

defines the element of the line i and the column j. If the number of lines is equal to the number of columns, that is, if $n = m$ we say that matrix A is a square matrix of dimension n. Other commands are given in Table 3.1.

MATLAB has an array editor. To illustrate its use we consider matrix A given above. We write again matrix A in the Command Window. This is shown in Figure 3.1 where we have also displayed the Command History window. In this window we see matrix A and in the Workspace we see the icon for the variable A specified as a 3×4 array. If we double click on this icon the Array Editor will open as a new window attached to MATLAB as shown in Figure 3.2. We can also open the array editor by typing in the Command Window

open A

There we can clearly see the element values which now can be either modified or deleted. All the elements have an imaginary part because at least one of them is complex, so all the elements are represented as complex ones. We can

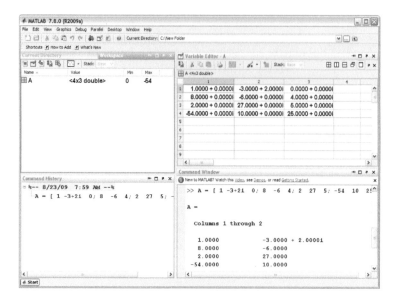

Figure 3.2: Array editor showing the elements of matrix **A**.

add new elements to matrix A. In this example we add a fourth row to A. The elements that are going to be added are

$$5 \qquad 10*i \qquad -27$$

We see in Figure 3.3 the elements added. Since we have an imaginary one, all the other elements are displayed as complex with zero imaginary part. In addition, we display in the Command Window the matrix A with the new row. If in the Command Window we write A, the matrix will be displayed as follows,

A =

1.0000	2.0000	3.0000	4.0000
5.0000	6.0000	7.0000	8.0000
9.0000	10.0000	11.0000	12.0000
57.0000	0 +10.0000i	27.0000	89.0000

3.2 Basic Operations with Matrices

The basic operations with matrices are addition, subtraction, multiplication, and division. The sum of two matrices can be carried out if both matrices are equal. of $n \times m$. For example for A, B and C given by

$$A = \begin{bmatrix} 3 & -7 & 3 \\ 4 & 9 & -1 \end{bmatrix}, \quad B = \begin{bmatrix} 4 & -12 & 7 \\ 8 & 6 & 2 \end{bmatrix}, \quad C = \begin{bmatrix} 2 & -7 \\ 4 & -8 \\ -9 & 6 \end{bmatrix}$$

The dimensions for each matrix are for A 2×3, for B 2×3 and for C 3×2. For addition or subtraction, it can only done for A and B because they have the same size 2×3 while C is of dimension 3×2. Then:

A+B
    ans =
    -1 5 9
    12 15 1

A-B
    ans =
    7 -19 -5
    -4 3 -3

B-A
    ans =
    -7 19 5
    4 -3 3

The multiplication of two matrices can be carried out if the number of columns of the first matrix is equal to the number of rows of the second matrix. In our matrices, the product can be performed for A and C, C and A, as well as that of B and C, C and B. These products are obtained with

A*C
    ans =
    -40 47
    53 -106

C*A
    ans =
    -22 -77 11
    -20 -100 16
    -3 117 -24

C*B
    ans =
    -64 -18 0
    -80 0 12
    84 -72 -51

B*C
```
    ans =
    -23 -26
    22 -92
```

If one of the matrices is a scalar, that is, it is a matrix of 1×1 dimension, the computations can also be done. For example, with $A1 = 2*A$ and $A2 = 2+A$, two new matrices are produced where all the elements from A multiply for two and they produce $A1$ and to all the elements from A we add 2 to obtain $A2$.

The inverse of a square matrix A is the matrix M such that

$$AM = MA = I$$

Where I is the identity matrix (the eye(n) matrix in MATLAB). The inverse of a square matrix A is obtained with

inv (A)

For the division of square matrices two possibilities exist, namely A/B and $A\backslash B$. A/B is the same as $A*$inv (B) while $A\backslash B$ is the same as inv$(A)*B$. While A/B and $A\backslash B$ can be used for any type of matrices, inv(A) can only be used for square matrices. For the matrices A and B given below we have that

A
```
    ans =
    0.3636          2.7273          1.4000
    0.7273          -0.5455         -0.4000
```

B
```
    ans =
    0.2500          1.2500          -0.2000
    0.3333          -0.1667         0.1000
```

A/B
```
    ans =
    1.9896          0.6391
    -0.0881         1.8845
```

For powers of matrices, there also exist two possibilities. If p is a scalar $A\hat{\ } p$ gives the the matrix A to the scalar p. On the other hand $p\hat{\ } A$ gives the scalar p to the A power. Some examples illustrate this for A and p given by

$$A = \begin{bmatrix} 25 & 49 \\ -3 & 57 \end{bmatrix} \qquad p = 2$$

we have

A^p
   115.6835   469.3423
   -28.7352   422.1927

A.^p
   1.0e+002 *
      1.2500   3.4300
    -0.0000 - 0.0520i   4.3034

p^A
   1.0e+009 *
     -0.3041   2.6826
     -0.1642   1.4477

p.^A
   1.0e+010 *
     0.0000 0.0425
     0.0000 1.0894

The point before the asterisk in **A.^p** and **p.^A** indicates that each element from the matrix A is to the power **p**. (This is different from **A^p** where the complete matrix is to the *p*th power). This operation is known as point exponential. (We also have the point multiplication and point division). The results are element-to-element operations for the matrix.

The instruction **poly** produces the characteristic polynomial $p(x)$ of a square matrix A which is defined as,

$$p(x) = det(A - xI) \tag{3.1}$$

where I is the identity matrix and **det** calculates the determinant of the matrix $A - xI$. For example, for the matrix A given by

A =
1   2   3
5   6   7
9   10 11

poly(A)
   ans =
   1.0000 -18.0000 -24.0000 0.0000

which means that the characteristic polynomial of the square matrix A is

$$p(x) = x^3 - 18x^2 - 24x \tag{3.2}$$

Table 3.2: Matrix Operations

diag (A)	Vector with the elements of the main diagonal.
inv A	Inverse of A. Only for square matrices.
A'	Transpose of A.
transpose (A)	Transpose of A.
det (A)	Determinant of A. Only if A is a square matrix.
rank(A)	Rank of A.
trace (A)	Sum of the elements of the main diagonal of A.
norm A	Norm of A.
A^c	Matrix A to the cth power.
A.^c	Each element of A to the cth power.
A/B	Same as $A*\text{inv}(B)$.
A\B	Same as $\text{inv}(A)*B$.
A.*B	Term to term product.
A./B	Term to term division.
poly(A)	Finds the characteristic polynomial of A.

Other operations with matrices are given in Table 3.1.

Some special matrices are:

ones (m,n)   Matrix of $m \times n$ dimension where each element has unity value.
ones (n)   Square matrix of n order where each element has unity value.
zeros (m,n)   Matrix with each element has zero value.
zeros (n)   Square matrix of order n where each element has zero value.
eye (n)   Square identity matrix of n order.
eye (m,n)   Matrix of $m \times n$ dimension with 1s in the main diagonal. The remaining elements are zeros.

Further information on matrices and arrays can be found in the MATLAB Help window which can be opened by clicking on the **Help** icon in the MATLAB tools bar. The help page for Matrices and Arrays is shown in Figure 3.3.

3.3   Vectors

A vector is a matrix with a single row called row vector, or with a single column, called column vector. Vectors obey matrix rules. As an example, the row vector given by:

x = [1 3 -7 4]

is a row vector of dimension 4, while

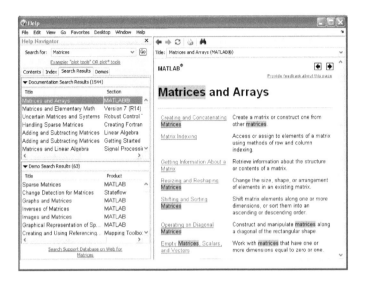

Figure 3.3: Help page for Matrices and Arrays.

$$y = [\ 9 \ ; \ 18 \ ; \ -5 \ ; \ 6 \ ; \ -7 \ ; \ 2 \ ; \ 4 \ ; \ 3 \ ; \ 11 \ ; \ 17 \ ]$$

is a column vector of dimension 10.

Note that a vector has the same structure in MATLAB as a polynomial does, according to the Chapter 2 definition for polynomials. Nevertheless, the operations that are carried out on them are completely different.

If we only desire to display on the Command Window some elements of a vector we use indexing. For example $y(k: m)$ specifies that we only wish to display from the kth to the mth elements. For example

```
y(2:4)
    ans =
    18
    -5
    6
```

and $y(i: j: k)$ gives the elements from the ith to the kth, but separated by j units. Then,

```
y(2 : 3 : 10)
    ans =
    18
    -7
    3
```

Table 3.3: Vectors operations. a and b are n dimensional vectors and c is a scalar

Operation	Result
a + c	$[\, a_1+c \ \ a_2+c \ \dots \ a_n+c \,]$
a * c	$[\, a_1 c \ \ a_2 c \ \dots \ a_n c \,]$
a + b	$[\, a_1+b_1 \ \ a_2+b_2 \ \dots \ a_n+b_n \,]$
a.b	$[\, a_1.b_1 \ \ a_2.b_2 \ \dots \ a_n.b_n \,]$
a./b	$[\, a_1/b_1 \ \ a_2/b_2 \ \dots \ a_n/b_n \,]$
a.\b	$[\, b_1/a_1 \ \ b_2/a_2 \ \dots \ b_n/a_n \,]$
a.^c	$[\, a_1\,\hat{}\,c \ \ a_2\,\hat{}\,c \ \dots \ a_n\,\hat{}\,c \,]$
c.^a	$[\, c\,\hat{}\,a_1 \ \ c\,\hat{}\,a_2 \ \dots \ c\,\hat{}\,a_n \,]$
a.^b	$[\, a_1\,\hat{}\,b_1 \ \ a_2\,\hat{}\,b_2 \ \dots \ a_n\,\hat{}\,b_n \,]$

The index j can be also negative. For example:

y(9 : -3 : 3)
    ans =
    11
    2
    -5

Some operations with vectors are given in Table 3.3.

The operations preceded by a point are term to term, as in the case of matrices.

If vectors a and b and scalar c are given by

$$a = \begin{bmatrix} 4 \\ 5 \\ 6 \end{bmatrix} \qquad b = \begin{bmatrix} -1 \\ -9 \\ 2 \end{bmatrix} \qquad c = 5$$

Then, we can perform the following operations:

a+c
    9
    10
    11

a*c
    20
    25
    30

a.*b
   8
   15
   -6

a.ˆb
   1.0e+002 *
       0.16000000000000
       2.43000000000000
       0.00166666666667

a.\b
       2.00000000000000
       1.66666666666667
       -0.16666666666667

c.ˆA
   144
   1728
   2985984

a.ˆb
   1.0e+002 *
       0.16000000000000
       2.43000000000000
       0.00166666666667

The vector product is only possible if we observe matrix multiplication rules. In this way, it is possible to multiply two vectors if the first one is a row vector with m columns and the second one is a column vector with m rows, or if we wish to multiply a column vector of n rows by a row vector of n columns. For example if we have the vectors

$$v_1 = \begin{bmatrix} 10 & 20 & 30 \end{bmatrix}, v_2 = \begin{bmatrix} a & b & c & d \end{bmatrix}, v_3 = \begin{bmatrix} 5 \\ 4 \\ 3 \\ 2 \\ 1 \end{bmatrix}, v_4 = \begin{bmatrix} 10 \\ 20 \\ 30 \end{bmatrix}$$

We will be able to perform only the following operations

```
syms a b c d
v1 = [10 20 30];
v2 = [a b c d];
v3 = [5; 4; 3; 2; 1];
```

v4 = [10;20;30];
v1*v4
10*10+20*20+30*30 = 1400

v4*v1
10*10, 10*20, 10*30
20*10, 20*20, 20*30
30*10, 30*20, 30*30

Any other operation that we perform will indicate an error because the dimensions of the vectors do not allow it.

In a similar way to the case of matrices, the basic operations that involve scalars can also be done. In this way, v-2 means that 2 is subtracted to each element of the vector v.

If we have a matrix A of dimension $n \times m$ and a vector b of dimension m, (that is, the number of columns of A is equal to the number of rows of b) we can multiply $A*b$. If A and b are

$$A = \begin{bmatrix} 3 & -2 & 0 \\ 4 & 9 & 17 \\ 2 & -4 & 9 \\ 6 & 2 & -5 \end{bmatrix} \qquad b = \begin{bmatrix} 3 \\ 2 \\ -7 \end{bmatrix}$$

Then A*b is

A = [3 -2 0; 4 9 17; 2 -4 9; 6 2 -5];
b = [3; 2; -7];
   A*b
   ans =
    5
   -89
   -65
   57

3.3.1   Norm of a Vector

The norm of a vector is a generalization of the length of a vector. The p-norm of an n-dimensional vector is defined by

$$\|x\|_p = \left(\sum_{k=1}^{n} |x_k|^p \right)^{1/p} \qquad (3.3)$$

We use

norm (x, p)

to obtain in MATLAB the norm of a vector. The default value for p is 2, corresponding to the Euclidean norm

$$\|x\|_2 = \left(\sum_{k=1}^{n} |x_k|^2\right)^{1/2} = \sqrt{x_1^2 + x_2^2 + ... + x_n^2} \tag{3.4}$$

also called 2-norm. The 1-norm is the sum of the absolute values of the vector components. That is

$$\|x\|_1 = \sum_{k=1}^{n} |x_k| = |x_1| + |x_2| + ... + |x_n| \tag{3.5}$$

The ∞-norm, also called Chebyshev norm or maximum norm is given by the greatest magnitude of the vector's components. That is,

$$\|x\|_\infty = max\{|x_k|\} \tag{3.6}$$

Thus, for x = [3 4 -2]', if we wish to find the 1-norm

norm (x, 1)
ans =
9

For the Euclidean norm **norm(x, 2)** or simply **norm(x)**

norm (x, 2)
ans =
5.3852

norm (x)
ans =
5.3852

And for the $\infty - norm$

norm (x, inf)
ans =
4

3.3.2   Vector Generation

A special type of vector is the one that we generate to evaluate a function in a given interval. For example, if it is desired to evaluate a function $f(x)$ in an interval $[a, b]$. If the distance between consecutive points is linear, we can use:

$$x = \text{linspace (a, b, n)}$$

This instruction generates a vector of n elements equally spaced between a and b. For example,

x = linspace(1, 10, 10)
  x = 1 2 3 4 5 6 7 8 9 10

To create a similar vector we can use

$$x = a : \text{increment} : b$$

Where a is the initial point and the following points are to a+increment, a+2*increment,..., a+k*increment,..., b. Depending on the value of the increment, the final value can be different from b. For example,

x = 3 : 0.7 : 6
x =
3.0000 3.7000 4.4000 5.1000 5.8000

To create a logarithmic spaced vector we use

$$x = \text{logspace (a , b , n)}$$

Where the first point is 10^a, the last point is 10^b and there are n points in the interval. For $a = 2$, $b = 5$ and $n = 4$ we have

x = logspace(2, 5, 4)
x =
100 1000 10000 100000

3.4   Dot and Cross Product

There are two important products involving vectors. They are the **dot product** and the **cross product**. The dot product result is a scalar and the cross product one is a vector.

3.4.1   Dot Product

The dot product of two vectors is also called scalar product and inner product. The dot product result is a scalar. For the vectors a and b given by

$$a = \begin{bmatrix} a_1 \\ a_2 \\ \vdots \\ a_n \end{bmatrix} , \qquad b = \begin{bmatrix} b_1 \\ b_2 \\ \vdots \\ b_n \end{bmatrix}$$

The dot product is defined by

$$a \cdot b = a_1b_1 + a_2b_2 + ... + a_nb_n \tag{3.7}$$

MATLAB finds the dot product with

dot (a , b)

For vectors a and b, we have

```
a = [ 2 ; 3 ; -2 ; 1 ];
b = [ 3 ; -8 ; 7 ; 4 ];
c = dot ( a , b )
   c =
        -28
```

The dot product can also be found with

$$a \cdot b = |a| \, |b| \, cos(\theta) \tag{3.8}$$

where θ is the angle between both vectors. We can readily see that if the vectors are orthogonal, then θ is $\pi/2$ and $cos(\pi/2) = 0$ and the dot product is equal to zero.

The dot product can also be evaluated if we transpose the first vector to make it a row vector and then we multiply it by the second column vector. That is,

$$a' * b \tag{3.9}$$

For the previously defined vectors we have

```
a'*b
   ans =
        -28
```

3.4.2 Cross Product

The cross product $a \times b$, also called vector product, is another vector orthogonal to the plane formed by the vectors a and b. The cross product can only be evaluated for three-dimensional vectors. MATLAB finds the cross product with

cross (a , b)

For vectors a and b given by

$$a = \begin{bmatrix} 9 \\ 6 \\ 23 \end{bmatrix} \qquad b = \begin{bmatrix} -27 \\ 11 \\ 12 \end{bmatrix}$$

we find the cross product with

    a = [9 ; 6 ;23];
    b = [-27 ; 11 ; 12];
    c = cross(a1, b1)
    c =
        -181
        -729
        261

The resulting vector has a magnitude (we evaluate the 2-norm) given by

    norm(c)
        ans =
        795.1874

3.5   Matrix and Vector Functions

It is possible to evaluate functions of matrices and vectors. When we request
to evaluate a function of a matrix, MATLAB makes the necessary calculations
(transparent to the user) to produce the result. Evaluating matrix functions
is not an easy task since it requires the evaluation of eigenvalues and eigenvectors, and in some cases generalized eigenvectors. This is required to transform
the matrix into the required form involving a modal matrix and a diagonal
one, or possible a quasidiagonal one. To show how easy it is to evaluate
matrix and vector functions in MATLAB, let us consider the row vector $x =$
$[2 \, 3 \, {-7}]$. To find the function $sin(x)$ we simply write in the Command Window:

    y = sin(x)
    y =
    0.90929742682568
    0.14112000805987
    -0.65698659871879

which is another vector whose components are the sine of each component of
the vector x.

    Functions defined in MATLAB, as well as user defined ones, can be used
with vectors and matrices. Thus, for any function $f(x)$, if A is a matrix, $f(A)$
is given by

$$f(A) = \begin{bmatrix} f(a_{11}) & f(a_{12}) & \cdots & f(a_{1n}) \\ f(a_{21}) & f(a_{22}) & \cdots & f(a_{2n}) \\ \vdots & \vdots & \vdots & \vdots \\ f(a_{n1}) & f(a_{n2}) & \cdots & f(a_{nn}) \end{bmatrix} \qquad (3.10)$$

And if v is a vector, then:

$$f(v) = \begin{bmatrix} f(v_1) \\ f(v_2) \\ \vdots \\ f(v_n) \end{bmatrix} \tag{3.11}$$

where v can be either a row vector or a column one. For example, for a polynomial defined by

$$p(x) = x^3 + 2 * x^2 - 7 * x + 17$$

where x can be a scalar, a vector, or a matrix. If D is a matrix given by

$$D = \begin{bmatrix} 2 & 5 \\ -4 & 9 \end{bmatrix}$$

then

```
D = [2 5; -4 9];
FD = D^3+2*D^2-7*D+17
    FD =
    -281 507 -375 405
```

3.6   Systems of Simultaneous Linear Equations

For a set of simultaneous linear equations such as

$$\begin{aligned} ax + by &= c \\ dx + ey &= f \end{aligned} \tag{3.12}$$

MATLAB can solve them, symbolically, with the instruction solve. For example, for the systems of equations given by Equation (3.12) we have:

```
[x, y] = solve('a*x+ b*y = c', 'd*x+e*y = f', 'x, y')
    x =
    -(-e*c+f*b)/(a*e-b*d)
    y =
    (a*f-c*d)/(a*e-b*d)
```

linsolve solves the same set of equations, but we have to write the system in matrix form. Thus, for the set given in Equation (3.12)

$$A = \begin{bmatrix} a & b \\ c & d \end{bmatrix} \qquad w = \begin{bmatrix} x \\ y \end{bmatrix} \qquad v = \begin{bmatrix} c \\ f \end{bmatrix}$$

This is equivalent to

$$Aw = v \tag{3.13}$$

This matrix equation can be solved with

```
syms a b c d e f x y
A = [a b; d e];
v = [c; f];
x = linsolve(A, v)
  x =
  [ -(-e*c+f*b)/(a*e-b*d)]
  [ (a*f-c*d)/(a*e-b*d)]
```

A set of simultaneous equations can also be solved to give a numerical solution by using matrix operations. To illustrate this, consider the set of simultaneous linear equations

$$
\begin{array}{rrrrl}
2a & -2b & & & = 5 \\
-2a & +6b & -2c & & = 0 \\
& -2b & +6c & -2d & = 0 \\
& & -2c & +8d & = 0
\end{array} \tag{3.14}
$$

That can be written in matrix form as

$$Ax = b \tag{3.15}$$

where

$$
A = \begin{bmatrix} 2 & -2 & 0 & 0 \\ -2 & 6 & -2 & 0 \\ 0 & -2 & 6 & -2 \\ 0 & 0 & -2 & 0 \end{bmatrix} \quad x = \begin{bmatrix} a \\ b \\ c \\ d \end{bmatrix} \quad b = \begin{bmatrix} 5 \\ 0 \\ 0 \\ 0 \end{bmatrix}
$$

To solve this system of equations, we first evaluate the inverse of A and then we multiply it by b:

```
A = [2 -2 0 0;-2 6 -2 0;0 -2 6 -2;0 0 -2 0];
b = [5; 0; 0; 0];
x = inv(A)*b
  x =
  3.7500
  1.2500
  0
  -1.2500
```

The same result can be obtained by the instruction

x = A\b
x =
3.7500
1.2500
0
-1.2500

which gives the same result but it is a little faster.

3.6.1   LU Factorization

LU factorization allows to write a square non-singular matrix A as the product of a lower triangular matrix L and an upper triangular matrix U, that is, we can write A as

$$A = LU \tag{3.16}$$

LU factorization can be performed with the instruction lu(A) and the instruction format is

$$[\ \mathbf{L}\ ,\ \mathbf{U}] = \mathsf{lu}\ (\mathbf{A})$$

For the matrix A given by

$$A = \begin{bmatrix} 1 & 2 & 3 & 4 \\ 5 & 6 & 7 & 8 \\ 9 & 10 & 11 & 12 \\ -3 & 4 & -6 & 29 \end{bmatrix}$$

A = [1 2 3 4; 5 6 7 8; 9 10 11 12; -3 4 -6 9]
A =
 1   2   3   4
 5   6   7   8
 9   10   11   12
-3   4   -6   9

[L, U] = lu(A)
L =
0.1111   -0.3333 0.0000 1.0000
0.5556   0.1333 1.0000 0
1.0000   0   0   0
-0.3333   1.0000 0   0

U =
9.0000 10.0000 11.0000 12.0000
0   3.3333 3.6667 13.0000
0   0   0.4000 -0.4000
0   0   0   7.0000

If matrix A is part of a system of linear equations in the form $Ax=b$, then the solution is

$$x = U\backslash(L\backslash b) \tag{3.17}$$

If $b = [2;\ 4;\ 0;\ 0]$ then for the matrix A given above we have

x = U\\(L\\b)
x =   ·
0.8571
-12.4286
10.2857
0.2857

3.7　Eigenvalues and Eigenvectors

A very important topic in matrix algebra concerns the way to find eigenvectors and eigenvalues. Eigenvectors are unique vectors that satisfy the following equation:

$$Ax = \lambda x \tag{3.18}$$

That is, when we multiply A by x we obtain the same vector x but multiplied by a constant λ called the eigenvalue of A corresponding to the eigenvector x. In other words, vector x only changes size but keeps the same direction. Eigenvalues can be either real or complex numbers. To find both eigenvectors and eigenvalues we use the instruction

[V, D] = eig(A)

The instruction **eig** returns two matrices. Matrix V has as columns the eigenvectors of A. Matrix D is a diagonal matrix whose elements of the main diagonal are the eigenvalues. As an example, let us consider matrix A given by

$$A = \begin{bmatrix} 4 & 0 & 1 & 0 \\ 2 & 2 & 3 & 0 \\ -1 & 0 & 2 & 0 \\ 4 & 0 & 1 & 2 \end{bmatrix}$$

Then, its eigenvectors and eigenvalues can be found with

A = [4 0 1 0;2 2 3 0;-1 0 2 0;4 0 1 2];
[V, D] = eig(A)
V =
0   0   0.2887   0.2887
1.0   0   -0.2887   -0.2887
0   0   -0.2887   -0.2887
0   1.0   0.8660   0.8660

D =
2 0 0 0
0 2 0 0
0 0 3 0
0 0 0 3

Thus we see that the eigenvalues are: $\lambda_1 = 2$, $\lambda_2 = 2$, $\lambda_3 = 3$, and $\lambda_4 = 3$. The corresponding eigenvectors are

$$v_1 = \begin{bmatrix} 0 \\ 1 \\ 0 \\ 0 \end{bmatrix}, v_2 = \begin{bmatrix} 0 \\ 0 \\ 0 \\ 1 \end{bmatrix}, v_3 = \begin{bmatrix} -0.2887 \\ -0.2887 \\ -0.2887 \\ 0.8660 \end{bmatrix}, v_4 = \begin{bmatrix} 0.2887 \\ -0.2887 \\ -0.2887 \\ 0.8660 \end{bmatrix}$$

Some matrices happen to have repeated eigenvalues and we cannot find all eigenvectors to be linearly independent. For example, if we try to repeat the procedure with matrix A given by

$$A = \begin{bmatrix} 2 & 2 & 1 & 0 \\ 0 & 2 & -1 & 0 \\ 0 & 0 & 2 & 0 \\ 0 & 0 & 0 & 1 \end{bmatrix}$$

we obtain

[V, D] = eig(A)
V =
1   -1   -1   -0.7071
0   0   0   0
0   0   0   0
0   0   0   0.7071
D =
2   0   0   0
0   2   0   0
0   0   2   0
0   0   0   1

We see that the matrix V does not have all of the column vectors linearly independent. In fact, the second and third columns are the same. This is a particular case arising when matrix A has one or several eigenvalues with multiplicity greater than unity. The procedure then is to find generalized eigenvectors. For a detailed treatment on generalized eigenvectors, readers can see reference [2]. The procedure to find generalized eigenvectors using MATLAB is

[V, J] = jordan (A)

where V is a matrix whose columns are the generalized eigenvectors and J is an almost diagonal matrix. The elements in the main diagonal are the eigenvalues and the elements in superdiagonal are equal to 1. We say that matrix J is in Jordan canonical form. For example, for matrix A given above

A = [2 2 1 1;0 2 -1 0;0 0 2 0;0 0 0 1]
[V,D] = jordan(A)
V =
-1.0000 1.0000 -1.0000 1.0000
0   0   0.5000 -0.2500
0   0   0   -0.5000
1.0000 0   0   0
V =
-1   1   -1   1
0   0   0.5 -0.25
0   0   0   -0.5
1   0   0   0

D =
1   0   0   0
0   2   1   0
0   0   2   1
0   0   0   2

We can see that in the superdiagonal of matrix D there are only 1s above the eigenvalues with multiplicity greater than unity.

3.8   Cell Arrays

Cell arrays are matrices where we can store different types of information in each cell. Each cell can be identified in the same way that matrix elements are identified. To show how a cell array can be created and used, let us store information about students in a cell array called **Student**. Each student will be assigned a column in the cell array. The information stored is: **Name**, **ID** number, **Semester**, **Homework grades**, **Exam grades**, **Year**. This is done in the following way:

Student 1,1 = 'Pablo Morales';
Student 1,2 = 19941027;
Student 1,3 = 'Fall 2012';
Student 1,4 = [10 9.8 8];
Student 1,5 = [9.9 8.9];
Student 1,6 = 2012;

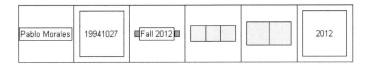

Figure 3.4: Cell array.

This technique to address arrays is called *content addressing*. We can "see" the array elements with

cellplot (Student)

to obtain Figure 3.4. To add a second student to the cell array we simply write

Student (2,1) = 'Laura Michele';
Student (2,2) = 19890510;
Student (2,3) = 'Fall 2012';
Student (2,4) = [10 10 7.8];
Student (2,5) = [10 9.6];
Student (2,6) = 2013;

Note that in this case we enclosed the data with parentheses. This manner to enter information is known as *cell indexing*. We can see a cell by typing:

Student(2, :)
ans =
'Laura Michele' [19890510] 'Fall 2012' [1x3 double] [1x2 double] [2013]

But if we write

Student2, :
ans =
Laura Michele
ans =
19890510
ans =
Fall 2012
ans =
10.0000 10.0000 7.8000
ans =
10.0000 9.6000

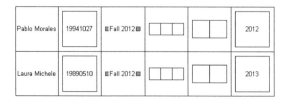

Figure 3.5: Cell array.

ans =
2013

We can print the complete cell again using

cellplot(Student)

and we obtain Figure 3.5.

3.9   Structures

Structures are arrays that can also store different types of information, but in this case each cell is identified by a field, that is, a name. Let us consider the cell arrays of the previous section. This information can be stored in a structure in the following way:

Student.Name = 'Pablo Morales';
Student.ID = 19941027;
Student.Semester = 'Fall 2012';
Student.Homework = [10 9.8 8];
Student.Exams = [9.9 8.9];
Student.Graduation = 2012;

In this structure we are defining six fields and each field is making reference to a different subject. All these fields are grouped under the variable **Student**. If we write this variable in the Command Window we obtain the information stored there.

Student
Student =
Name: 'Pablo Morales'
ID: 19941027
Semester: 'Fall 2012'

Homework: [10 9.8000 8]
Exams: [9.9000 8.9000]
Graduation: 2012

If we only wish to see a specific field we only write **Student** and the file name separated by a dot. For example if we wish to see the data for the field **Semester** we write

Student.Semester
ans =
Fall 2012

To change or to add data we proceed as in the case of matrices. Thus, if we wish to change the **Homework** data to 10 9.9 9.8, then we only write

Student.Homework = **[10 9.9 9.8];**

To add a third exam we concatenate the new data to the previous data, as in

Student.Exams = **[Student.Exams, 9.7];**

To add a second student we write

Student(2).Name = **'Laura Michele'**

The variable **Student** now has two elements. The first element is **Student(1)** and the second element is **Student(2)**. We can add data to the new **Student** as in

Student (2).ID = **19890510;**
Student (2).Semester = **'Fall 2012';**
Student (2).Homework = **[10 10 7.8];**
Student (2).Exams = **[10 9.6];**
Student (2).Graduation = **2013;**

We can add now as many students as we need by following the same procedure.

3.10   Concluding Remarks

This chapter has covered the way MATLAB handles matrices and vectors and how some fundamental topics in Linear Algebra can be solved with MATLAB. The basic operations that can be done with them were described and some examples show them. Special matrices for data, called structures, were described and some examples were presented. Systems of linear equations can be solved with MATLAB in a straightforward way. Topics such as norm, dot and vectro products were treated in detail as well as the concept of eigenvalues and eigenvectors. The case of repeated eigenvectors was covered too.

3.11   Exercises

Section 3.1

3.1 Given the matrix

$$A = \begin{bmatrix} 14 & 1 & 27 \\ 27 & 54 & 2 \\ 10 & 5 & 9 \\ 89 & 50 & 19 \end{bmatrix}$$

Write only: (a) the second row, (b) the third column, (c) the trace.
3.2 From the matrix A in Exercise 3.1, obtain the submatrix which contains the elements from columns 2 and 3 and rows 1 and 4.

Section 3.2

For the examples in this section, the matrices are defined by

$$A = \begin{bmatrix} 1 & 7 & 2 & -5 \\ -1 & -1 & 8 & 4 \\ 2 & 4 & 5 & 3 \\ 6 & 9 & -5 & 1 \end{bmatrix}, \quad B = \begin{bmatrix} 4 & -3 \\ 8 & 0 \\ 1 & -5 \\ -6 & 2 \end{bmatrix},$$

$$C = \begin{bmatrix} 1 & 0 & 2 & 3 \\ -1 & 1 & 0 & 4 \\ 2 & 1 & -1 & 3 \\ -1 & 0 & 5 & 7 \end{bmatrix}, \quad D = \begin{bmatrix} 3 & -4 & -2 & -1 \\ 7 & 8 & -6 & 9 \end{bmatrix}.$$

3.3 Determine the dimensions of the matrices A, B, C and D.
3.4 Obtain the product of A and B.
3.5 Obtain the product of A and C.
3.6 Check if the products $A*D$ and $D*A$ and obtain them.
3.7 Evaluate A-C.
3.8 Obtain the inverse for the matrices A and C.
3.9 Evaluate A/B and B/A.
3.10 Evaluate A^3 and $A.^3$.
3.11 Find the matrices $2\hat{\ }A$ and $2.\hat{\ }A$.
3.12 Obtain the transpose of A and B.
3.13 Obtain the determinant for the matrices C and A.
3.14 Find the trace for D and C.

3.15 Obtain $A.*B$ and compare it with the result of $B.*A$.

Section 3.3

3.16 For the row vector $v=[1\ \text{-}3\ 5\ 8]$, find the 1-norm and the 2-norm.
3.17 Find the dot and cross products for the vectors $v=[2\ 4\ \text{-}9]$, $w=[\text{-}4\ 6\ \text{-}10]$. Also, for the scalar $a=2$. Evaluate
    (a) $v.*w$, (b) $2.^{v}$, (c) v^2.
3.18 Generate an interval with linearly spaced points which starts at $x=3$, ends at $x=17$, and with 101 points.
3.19 Generate an interval with logarithmically spaced points which starts at $x=200$ and ends at $x=300$ and having 50 points.

Section 3.4

3.20 For the vectors $a=[3\ 4\ 6\ 8]$, $w=[\text{-}1\ 2\ 0\ \text{-}8]$ obtain the dot product.
3.21 Find the angle between vectors.
3.22 Obtain the dot and cross products for vectors $a=[\text{-}4\ 0\ \text{-}3]$, $b=[2\ 7\ \text{-}1]$.
3.23 Evaluate the dot and cross products for pairs of vectors:
    $\hat{\imath}=[1\ 0\ 0]$, $\hat{\jmath}=[0\ 1\ 0]$, and $\hat{k}=[0\ 0\ 1]$.
3.24 Evaluate the cross product for vectors $a=[\text{-}4\ 0\ \text{-}3]$, $w=[2\ 7\ \text{-}1]$.

Section 3.5

3.25 Evaluate the function $\tan(A)$ for the matrix

$$A = \begin{bmatrix} \pi & 0 & -2 & 5 \\ -1 & -1 & 8 & 2 \\ -2 & 6 & 5 & 0 \\ 9 & 0 & -3 & 7 \end{bmatrix}$$

3.26 For vectors $a=[8\ \text{-}3\ \text{-}7]$, $b=[0\ \text{-}3\ 2]$, $c=[\text{-}4\ 5\ 1]$, evaluate:
    (a) $\sin(a)$,
    (b) $a\hat{\ }2$,
    (c) $\sqrt{b}$,
    (d) if $p(x)$ is a polynomial given by $p(x)=3x^2-1$, evaluate $p(c)$.
3.27 Evaluate $p(A)$ where $p(x)=-7x^3-8x^2-x+11$ and A is given by

$$A = \begin{bmatrix} 5 & 1 & 6 & -3 \\ -7 & -5 & 2 & 8 \\ 0 & -6 & 3 & 1 \\ 6 & 10 & -13 & 4 \end{bmatrix}$$

Section 3.6

3.28 Use solve to find a solution for the system of equations

$$3x - 7y + 9z + 3w = 5$$
$$-2x + 5y - 4z + 8w = -1$$
$$8x - 4y + z + 2w = 6$$
$$-x - 3y + 4z - 5w = 0$$

3.29 Repeat Exercise 3.28 using linsolve.
3.30 Repeat Exercise 3.28 by evaluating first the inverse of the system matrix and then multiplying it by the vector of independent coefficients.
3.31 Repeat Exercise 3.28 using the instruction rref.

Section 3.7

3.32 Obtain the LU factorization for the matrix

$$A = \begin{bmatrix} 12 & 8 & 21 \\ 6 & -2 & -4 \\ 3 & -3 & -6 \end{bmatrix}$$

3.33 Obtain the LU factorization for the matrix

$$A = \begin{bmatrix} 2 & 8 & 0 \\ 2 & 2 & -3 \\ 1 & 2 & 7 \end{bmatrix}$$

Section 3.8

3.34 Obtain the eigenvalues and eigenvectors for the matrix

$$A = \begin{bmatrix} 2 & 1 & 2 \\ 2 & 2 & -2 \\ 3 & 1 & 1 \end{bmatrix}$$

3.35 Obtain the eigenvalues and eigenvectors for the matrix

$$A = \begin{bmatrix} 5 & 1 & -1 \\ 0 & 5 & 2 \\ 0 & 0 & 5 \end{bmatrix}$$

3.36 Obtain the eigenvalues and eigenvectors for the matrix

$$A = \begin{bmatrix} 2 & 1 & 0 & -1 \\ 0 & 2 & 1 & 1 \\ 0 & 0 & 2 & 0 \\ 0 & 0 & -0 & 2 \end{bmatrix}$$

Sections 3.9 and 3.10

3.37 Create a structure with the following data:

Player's Name	Team	Number	Position	Goals
Ronaldo	Brazil	10	Forward	17
Pelé	Brazil	10	Forward	18
Maradona	Argentina	10	Forward	14

3.38 To the structure created in Exercise 3.37, add the following data: Beckenbauer, Germany, 4, Midfielder, 3.

3.39 Repeat Exercises 3.37 and 3.38 but using cell arrays.

3.12 References

[1] D. C. Hanselman, Mastering MATLAB 7, Prentice Hall, Inc., Piscataway, NJ, 2004.

[2] B. Kolman and D. Hill, Elementary Linear Algebra with Applications, 9th Ed., Prentice Hall, Inc., Piscataway, NJ, 2008.

[3] I.M. Gelfand and A. Shen, Algebra, Birkhäuser, Boston, MA, 2003.

[4] H. Moore, MATLAB for Engineers, 2nd Edition, Prentice Hall, Inc., Piscataway, NJ, 2008.

[5] R. Bronson, Matrix Methods, An Introduction, Academic Press, Inc., N.Y, 2008.

[6] M. E. Herniter, Programming in MATLAB, Brooks/Cole Thomson, Pacific Grove, CA, 2001.

Chapter 4

Calculus

4.1   Introduction

MATLAB allows users to perform a number of operations on functions, such as differentiation and integration. We can also find limits and check for continuity at a point. These operations can be done in a numerical way or in a symbolic one. In addition, we can solve differential equations with great simplicity. Such simplicity is an indication of how powerful is MATLAB to solve complicated tasks. In the case of operations involving symbolic functions, the result is also a symbolic function. However, the resulting function can be evaluated numerically or plotted to see its behavior. The availability of symbolic computations is due to the existence of the Symbolic Toolbox available from The MathWorks, Inc. This toolbox has to be installed to be able to perform symbolic computations.

The chapter is organized as follows: We start with limits of sequences and functions. We continue with differentiation and integration. Finally, we show how to solve differential equations, both numerically and symbolically.

4.2   Limits of Functions

Limit calculation for sequences and functions is an important topic in calculus. The limit of a function $f(x)$ as x approaches x_o is given by

$$\lim_{x \to x_0} f(x) \tag{4.1}$$

MATLAB provides the instruction limit to find the limit of a function. The format is

$$\text{limit } (f, x, x_0)$$

95

For example, for the function $f(x) = x^2$, the limit as x approaches 3 can be found with

```
syms x
limit ( x^2 , x , 3)
    ans =
    9
```

And the limit of $g(x) = \sqrt{x^2 + 1}$ as x approaches ∞ is found with

```
syms x
limit ( sqrt ( x^2+1) , x , inf )
    ans =
    inf
```

In the definition of limit there is no restriction as to how x should approach x_0. However, in some cases we need to specify the way x approaches x_0. For real variables, x can approach x_0 from the left or from the right. These limits are called lateral limits. We can find the limit from the right as

$$\text{limit (f, x, a ,'right')}$$

and for the limit from the left the instruction is

$$\text{limit (f, x , a , 'left')}$$

For example, if the function $f(x)$ is

$$f(x) = \frac{|x - 3|}{x - 3}$$

If we evaluate the left and the right limits we obtain

```
syms x
f = abs(x-3)/(x-3);
limit(f, x, 3, 'left')

    ans =
    -1
limit(f, x, 3, 'right')
    ans =
    1
```

If the limit of a function exists, it is unique. Thus, in the case of this function the limit as x approaches 3 does not exist.

The derivative of $f(x)$ at $x = x_0$ is defined as the limit

$$\lim_{x \to x_0} \frac{f(x) - f(x_0)}{x - x_0} \tag{4.2}$$

For the function $f(x)$ defined by

$$f(x) = \frac{1}{x - 2}$$

The derivative is

$$\frac{df(x)}{dx} = \frac{1}{(x - 2)^2}$$

Using limits, we can find the derivative at x_o as

```
syms xo
limit((1/(x-2)-1/(x0-2))/(x-x0), x, x0)
    ans =
    -1/(x0-2)^2
```

which is the expected result.

Limits can be nested. Nested limits appear frequently when we have functions of several variables. For example, in the function

$$f(x, y) = 2xy - x^2$$

We can find the limits as x approaches 2 and as y approaches 3 as

```
f = 2*x*y-x^2;
limit(limit(f, x, 3), y, 2)
    ans =
    3
```

We can nest as many limits as we wish but it is recommended not to nest more than two limits.

4.3 Limits of Sequences

A sequence is a function whose domain is the set of positive integers. We usually write a sequence as $a = \{a_n\}$ and it is understood that the value of n goes from 1 to infinity. We say that a sequence converges to a number A as n approaches if for each $\epsilon > 0$ there exists a positive integer N such that for all $n > N$ we have $|a_n - A| < \epsilon$. The real number A is called the limit of $\{a_n\}$ as n approaches ∞. If a sequence does not converge we say that it diverges.

We can also find limits of sequences $\{a_n\}$ with MATLAB. Let us consider the sequence

$$a_n = \sqrt[n]{\frac{1+n}{n^2}}$$

It is desired to find

$$\lim_{n\to\infty} a_n = \lim_{n\to\infty} \sqrt[n]{\frac{1+n}{n^2}}$$

In MATLAB we use

```
syms n
limit ( ( ( 1+n) / n^2 )^( 1/n ) , inf )
    ans =
    1
```

As another example, let us consider the sequence

$$a_n = \frac{n^2 + n}{n^2 - n}$$

Whose limit we can find with

```
an = (n^2+n)/(n^2-n);
limit(an, n, inf)
    ans =
    1
```

Now, if the sequence is

$$a_n = \frac{3n^2 + 4n}{2n - 1}$$

We can look for the limit as

```
an = (3*n^2+4*n)/(2*n-1);
limit(an, n, inf)
    ans =
Inf
```

Which indicates that this sequence diverges.

4.4 Continuity

A function $f(x)$ is continuous at a point $x = a$ if

$$\lim_{x \to a} f(x) = f(a) \tag{4.3}$$

If the limit does not exist, we say that the function is discontinuous at that point. If the limit exists but it is not equal to $f(a)$ it is said that the function has a discontinuity at $x = a$. Let us consider the functions

$$f(x) = \frac{\sin x}{x}, g(x) = \sin \frac{1}{x}$$

The limit of $f(x)$ as x approaches $x = a$ can be found with

```
syms x a
limit ( sin ( x ) / x , x , a)
    ans =
    sin(a)/a
```

From this result we might think that the limit does not exist at $x = 0$. But when we find the limit as x approaches $x = 0$ we get

```
limit( sin(x)/x , x , 0 )
    ans =
    1
```

For $x = a$

```
limit( sin(1/x), x , a )
    ans =
    sin(1/a)
```

But for $x = 0$

```
limit( sin(1/x) , x , 0 )

    ans =
    -1 .. 1
```

Which indicates that the limit is in the interval [-1, 1]. If we now evaluate the lateral limits we get

```
limit ( sin ( 1/x ), x , 0, 'left' )
    ans =
    -1 .. 1
```

limit (sin (1/x), x , 0 , 'right')
   ans =
   -1 .. 1

Which indicates that this function is discontinuous at $x = 0$.

Now, let us consider the function $f(x, y)$

$$f(x, y) = \frac{xy}{x^2 + y^2}$$

This domain of this function is the whole plane except for the point (0,0). To find the limit as (x, y) approaches a point (x_0, y_0) in the plane, we can approach this point from any direction in the plane. For example, to find the limit we can use

syms x y
limit (limit (x*y / (x^2 + y^2) , x , 0) , y , 0)
   ans =
   0

limit (limit (x*y / (x^2 + y^2) , y , 0) , x , 0)
   ans =
   0

Indicating that this function is continuous at the point (0, 0). We can also use a set of straight lines passing through the origin. For example, the set of lines

$$y = mx$$

where m is the slope of the lines that pass through the origin. Using this equation, the function $f(x, y)$ becomes

$$f(x) = \frac{mx^2}{x^2 + mx}$$

Then the limit is

syms x m
limit((m*x^2)/(x^2 + m*x), x , 0)
   ans =
   0

We can use any family of curves. For example, for the family of paraboles

$$y^2 = mx$$

Then, $f(x, y)$ now becomes

$$f(x, y) = \frac{m^{1/2}x^{3/2}}{x^2 + mx}$$

And the limit is

```
limit ( m^(1/2)*x^(3/2) )/(x^2+m*x), x , 0 )
    ans =
    0
```

Which again proves that the function is continuous at the origin.

4.5 Derivatives

Let a function $f(x)$ be defined at any point in the interval $[a, b]$. The derivative of $f(x)$ at $x = x_0$ is defined as

$$f'(x) = \lim_{x \to x_0} \frac{f(x) - f(x_0)}{x - x_0} \tag{4.4}$$

If the limit exists, the function $f(x)$ is called differentiable at the point $x = x_0$. If $f(x)$ is differentiable at a point, the function is continuous at that point.

MATLAB can evaluate derivatives in a symbolic manner using the instruction diff. To see how it works, let us consider the function

$$f(x) = -3x^3$$

We know that its derivative is

$$f'(x) = -9x^2$$

Now, using MATLAB we get

```
f = - 3*x^3;
diff(f)
    ans =
    -9*x^2
```

To find the second derivative of $f(x)$ we use

```
diff(f, 2)

    ans =

    -18*x
```

In the case of function of several variables, we must indicate with respect to which variable we wish to find the derivative. For example, for the function $f(x) = 2x^3y + 3x/y$, we wish to differentiate first with respect to y and then with respect to x. Then, we use

```
syms x y
f = 2*x^3*y+ 3*x/y;
f1 = diff(f, y)
   f1 =
   2*x^3-3*x/y^2
f2 = diff(f1, x)
   f2 =
   6*x^2-3/y^2
```

If we do not indicate the variable with respect to which we wish to differentiate, MATLAB differentiates with respect to the last few letters in the alphabet. Thus, in the case of the function $f(x) = axy + y$, MATLAB finds

```
f(x) = a*x*y+y;
f1 = diff(f)
   f1 =
   a*y
```

which is the derivative with respect to x

```
f1 = diff(f,x)
   f1 =
   a*y
```

and this is the same result.

```
f1 = diff(f,y)

   f1 =

   a*x+1
```

```
f1 = diff(f, a)

   f1 =

   x*y
```

Now, let us see the following results:

```
f = a*x*y+y;
f1 = diff(f)
   f1 =
   a*y
```

which is the derivative with respect to x. Now, the resulting function is independent of x. Then,

```
f1 = diff(f1)
   f1 =
   a
```

Since f1 only has as variables y and a, the derivative is taken with respect to y which belongs to the set of the last few letters in the alphabet. Now, the only variable is a and the derivative is taken with respect to a, as in

```
f1 = diff(f1)
   f1 =
   1
```

As another example, if $f(x) = x^n$, the derivative with respect to x is

```
syms x n
f = x^n;
diff ( f )
   ans =
   x^n*n / x
```

If we wish to differentiate with respect to n we must explicitly write it as

```
diff(f, n)
   ans =
   x^n*log ( x )
```

Note that in the case of several variables, the derivative that we obtain with diff is the partial derivative

$$\text{diff}(f, x) = \frac{\partial f}{\partial x} \quad and \quad \text{diff}(f, y) = \frac{\partial f}{\partial y} \tag{4.5}$$

4.6   Integration

Integration is defined for continuous functions. Thus if a function is continuous in a region, then the integral of $f(x)$ is defined as another function $F(x)$ such that

$$F(x) = \int f(x)dx \qquad (4.6)$$

The function $F(x)$ satisfies the condition

$$f(x) = \frac{dF(x)}{dx} \qquad (4.7)$$

F is also called the antiderivative of $f(x)$.

MATLAB calculates the integral of a function $f(x)$ with the instruction int(f). For example, for the function $f(x) = 2x$, the integral is given by

```
int('2*x')
    ans =
    x^2
```

In the case of functions of several variables we have to indicate with respect to which variable we wish to integrate. For example, for the function $f(x) = 2x^y$, its integral with respect to y is given by

```
int( '2*x^2*y', 'y')
    ans =
    x^2*y^2
```

The definite integral is evaluated by giving the limits after the function specification in the instruction int. Thus, for the function $2x^2 + 5x$, the integral between the limits 2 and 3,

```
int('2*x^2+5*x', 2, 3)
    ans =
    151/6
```

Improper integrals such as

$$\int_0^\infty \frac{e^x \sin x}{x} dx$$

Can be evaluated in the same way as definite integrals, thus

```
syms x
f = ( exp (-x )*sin ( x ) ) / x;
int ( f , 0 , inf )

    ans =
    1/4*pi
```

Some integrals do not converge. For example, the function $f(x) = \tan(x)$ has a singular point at $x = \pi/2$, thus, to evaluate

$$\int_0^{\pi/2} \tan x\, dx$$

We use,

```
syms x b
f = tan ( x );
int ( f , x , 0 , pi/2)
    ans =
    inf
```

and then we conclude that the integral does not exist.

Sometimes, MATLAB is unable to find a closed form solution. In those cases we have to use a numerical solution. There are several instructions to integrate numerically. One of them is the instruction quad. It uses a Simpson rule to find the integral. The format is the same as in the symbolic instruction int. For example, for the function $f(x) = \sin(x)/x$, the integral

$$\int_0^{\pi/2} \sin x\, dx$$

can be found with

```
quad('sin(x)', pi/2, pi)
    ans =
    1.0000
```

The function $f(x)$ can be defined in an m-file

```
function y = mifuncion(x)
y = sin(x);
```

and then,

```
quad(@mifuncion, pi/2, pi)
    ans =
    1.0000
```

With this instruction we can find line integrals. For example, the line integral

$$\int \frac{s+1}{s-1-2i}\, ds$$

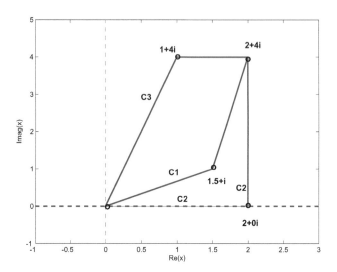

Figure 4.1: Integration paths.

We see that the function to be integrated has a singular point at $s = 1+2i$ and thus the path must not pass through it. We wish to find the value of the integral from the point 0 to 2 +4i. In Figure 4.1 we see three possible paths. Because the function is analytic everywhere except at the singular point, the integral along any one of these paths would give the same value. For path C_1 we first integrate along the segment that starts at the origin and ends at 1.5+i and then from the point 1.5+i to the end point in the path 2+ 4i. We do this with

quad('(s+1)./(s-1-2*i)', 0, 1.5+i)+quad('(s+1)./(s-1-2i)', 1.5+i, 2+4i)
    ans =
    -4.2832 +10.2832

We now integrate along C_2

quad('(s+1)./(s-1-2i)', 0, 2)+ quad('(s+1)./(s-1-2*i)', 2, 2 + 4i)
    ans =
    -4.2832 +10.2832i

We obtain the same result if we integrate along path C_3.
Let us consider now the function $|s|$. We know this function is not analytic on the s-plane and, therefore, its integral depends upon the path we choose for the integral. On C_1 we obtain

quad('abs(s)', 0, 1.5+i)+ quad('abs(s)', 1.5+i, 2 + 4*i)
    ans =
    2.8911 +10.1356i

On C_2 we get

quad('abs(s)', 0, 2)+ quad('abs(s)', 2, 2 + 4i)
    ans =
    2.0000 +11.8315i

and along C_3 we obtain

quad('abs(s)', 0, 1+4i)+ quad('abs(s)', 1+4i, 2 + 4i)
    ans =
    6.3421 + 8.2462i

As we see, the integral in the three cases yields a different value.

The instruction quad employs the Simpson method for the evaluation of the integral. Another method available for integral evaluation is the trapezoidal method which is available with the instruction trapz.

4.7 Series

Let us consider the sequence $\{a_n\}$ where the elements a_n can be numbers or functions. The formal sum of them

$$a_1 + a_2 + a_3 + \dots + a_n + \dots = \sum_{n=1}^{\infty} a_n$$

is called an infinite series. The partial sums are defined by

$$S_n = \sum_{n=1}^{\infty} a_k$$

The partial sums are a sequence. If the limit

$$\lim_{n \to \infty} S_n = S$$

exists, then S is called the sum or the limit of the series. In this case, we say that the series converges and

$$S = \sum_{n=1}^{\infty} a_k$$

If the limit does not exist we say that the series diverges.

MATLAB can find the sum of a series of the form

$$S = \sum_{n=0}^{\infty} f_n \tag{4.8}$$

with the instruction symsum that has the format

$$S= \mathsf{symsum(f,\ n,\ a,\ b)}$$

where f_n is the nth term in the series, n is the index, and a and b are the lower and upper limits, respectively. For example, for the power series

$$S = \sum_{n=0}^{\infty} x^n \tag{4.9}$$

We can find the sum with

```
syms x n
f = x^n;
sum = symsum ( f , n , 0 , inf )
    sum =
    -1/(x-1)
```

We recall from calculus that this series converges if $|x|<1$. Thus, although MATLAB has found a general solution for the sum, we must be careful about the conditions for convergence. We can check this by substituting values for x, say $x = 0.5$ and $x = 2$. We have then,

```
f = .5^n;
sum = symsum ( f , n , 0 , inf )
    sum =
    2
```

```
f = 2^n;
sum = symsum ( f , n , 0 , inf )
    sum =
    Inf
```

In the first case we see that the series converges, but for $x = 2$ the series diverges.

As another example, let us consider the series,

$$\sum_{n-0}^{\infty} \frac{2n+3}{(n+1)(n+2)}$$

Using MATLAB we can find the sum as

```
f = (-1)^n*(2*n+3)/(n+1)/(n+2);
sum = symsum ( f , n , 0 , inf )
   sum =
   1
```

Thus, the series converges.

We know that the harmonic series diverges. This series is given by

$$a_n = \frac{1}{n}$$

We can see this with

```
an = 1/n;
sum = symsum ( an , n , 1 , inf )
   sum =
   Inf
```

4.8 Differential Equations

A differential equation is an equation involving derivatives. They arise in many fields of physics, mathematics, engineering, economy, finances, etc. Examples of phenomena described by differential equations are Newton's second law of mechanics, Hooke's law, electric circuits, a simple pendulum, cash flow, and birth rate are a few examples of systems that can be described by differential equations.

An nth order ordinary differential equation can be described by the equation

$$\frac{d^n y(t)}{dt^n} = f\left(y(t), \frac{dy(t)}{dt}, \frac{d^2 y(t)}{dt^2}, \dots, \frac{d^{n-1} y(t)}{dt^{n-1}}, t \right) \tag{4.10}$$

This equation can be solved by MATLAB. The representation of a derivative in MATLAB, for the symbolic solution of differential equations is Dy for the first derivative, D2y for the second derivative, and for the nth derivative

$$\frac{d^n y(t)}{dt^n} \rightarrow \text{Dny} \tag{4.11}$$

Then, the following differential equation is described in MATLAB notation as

$$\frac{d^2 y(t)}{dt^2} = 1 + \frac{dy(t)}{dt} + y^2 t \rightarrow \text{D2y(t)=1+Dy(t)+y^2*t} \tag{4.12}$$

A differential equation can be solved symbolically with the instruction dsolve. As an example, the differential equation

$$\frac{d^2y(t)}{dt^2} + y(t) = 4$$

can be solved in MATLAB with

dsolve('D2y+y = 4 ')
ans =
4+C1*sin (t)+C2*cos (t)

C1 and C2 are constants which can be found from the initial conditions. If the initial conditions are

$$y(0) = 1 \quad y'(0) = 0$$

Then we add them to the instruction dsolve as

dsolve ('D2y+y = 4', 'y(0) = 1', 'Dy(0) = 0')
ans =
4-3*cos (t)

Note that the independent variable is not explicitly defined, but MATLAB has as the default independent variable t. We may, however, assign the independent variable. For example, if we want x as the independent variable, then we use

dsolve('D2y+y = 4' , 'y(0) = 1' , 'Dy(0) = 0' , 'x')
ans =
4-3*cos(x)

For the second-order differential equation

$$\frac{d^2y(t)}{dt^2} + a\frac{dy(t)}{dt} + by(t) = c \tag{4.13}$$

The instruction dsolve is

y = dsolve (' D2y+a*Dy+ b*y = c')
y =
1/b*c+C1*exp(-1/2*(a+(a^2-4*b) ^(1/2))*t)+...

C2*exp(-1/2*(a-(a^2-4*b)^(1/2))*t)

We use the instruction pretty to display the solution in a more readable format,

pretty(y)

$$c/b + C1 \exp\left(-1/2(a + (a-4b)^{1/2})t\right) + C2 \exp\left(-1/2(a - (a-4b)^{1/2})t\right)$$

For the systems of linear differential equations

$$\frac{dx(t)}{dt} = 3x(t) + 4y(t)$$

$$\frac{dy(t)}{dt} = -7x(t) + 6y(t)$$

We can also use the instruction dsolve as

[x, y] = dsolve('Dx = 3*x+4*y','Dy = -7*x+6*y','x(0) = 2','y(0) = 1')

x =
-1/103*exp(9/2*t)*(-206*cos(1/2*t*103^(1/2))....
    25cm -2*103^(1/2)*sin(1/2*t*103^(1/2)))

y =
-1/103*exp(9/2*t)*(25*103^(1/2)*sin(1/2*t*103^(1/2))...
    -103*cos(1/2*t*103^(1/2)))

Again, we use pretty to display the solution in a more readable format

pretty(x)

1/103 exp(9/2 t) (206 cos(1/2 t 103^{1/2}) + 2 103^{1/2} sin(1/2 t 103^{1/2}))

pretty(y)

1/103 exp(9/2 t) (-25 103^{1/2} sin(1/2 t 103^{1/2}) + 103 cos(1/2 t 103^{1/2}))

A technique for solving linear differential equations is by separation of variables. To see how we can use dsolve we consider the differential equation

$$y \sin t \, d - (1 + y^2) dy = 0$$

That can be rewritten as

$$\sin t \, dt = \frac{1 + y^2}{y} dy$$

We now use dsolve on each term and then equate both results. For the left-hand term we have

syms t
int (sin(t))
   ans =
   -cos(t)

And for the right-hand term we have

syms y
int ((1+y^2) / y)
   ans =
   1/2*y^2+log(y)

We now equate both results and use the instruction solve to find the solution of the differential equation,

solution = solve('-cos(t) = 1/2*y^2+log(y)')
   solution =
   exp(-1/2*lambertw(exp(-2*cos(t)))-cos(t))

where lambertw is the Lambert W function. The Lambert W function is the inverse function of

$$f(W) = We^W$$

4.8.1 Numerical Solution of Differential Equations

MATLAB provides instructions to numerically solve a differential equation. Two of the most used ones are ode23 and ode45. The instruction ode23 uses Runge-Kutta second- and third-order methods. With ode23, a second-order method is used first and then a third-order method. A similar description applies for ode45 but in this case fourth- and fifth-order methods are used. The difference between the two methods is that ode23 uses more crude tolerances. Both methods are one step solvers. The format for these instructions is

$$[t, y] = \text{ode23 (F, [t\_initial, t\_final], Yo)}$$
$$[t, y] = \text{ode45 (F, [t\_initial, t\_final], Yo)}$$

Here, F is a string of text and it indicates where the differential equation is defined, usually an m-file. t_initial and t_final are the initial and final times for the simulation and Yo is the initial condition. As an example, let us consider the following differential equation

$$\frac{dy}{dx} = -2yt$$

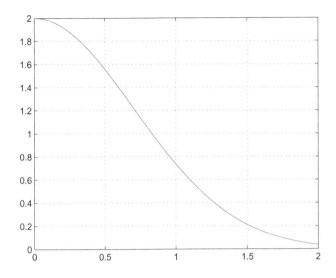

Figure 4.2: Solution of the differential equation $dy/dt = -2yt$.

We define this equation in an m-file dy.m and then we save it in the MATLAB directory. The m-file is

```
function yderivative = dy (t, y)
% This is file dy.m
yderivative = -2*y*t;
```

we now solve the differential equation with ode23 with

```
y0 = 2;
[ t , y ] = ode45 ( 'dy', [0, 2] , y0 );
plot ( t , y )
grid
```

The result is plotted in Figure 4.2. The exact solution is $y(t) = 2 * \exp(-t^2)$. If we plot this exact solution in the same plot we find that the match is exact. We can plot the exact solution with

```
t = 0: 0.1: 2;
y1 = 2*exp( -t.^2);
plot(t, y1)
```

4.9   Concluding Remarks

MATLAB can also work calculus. In this chapter we work with symbolic techniques to differentiate, integrate, find limits of sequences and series, and to solve differential equations. However, it can also perform numerical computations. We presented examples for line integral evaluation and numerical solution of differential equations.

4.10   Exercises

Section 4.1

4.1 Find the limit for the function $sin(x)$ as x approaches 0.
4.2 Find the limit for the function

$$f(x, y) = \frac{x - 1}{x^2 - 1}$$

as x approaches 1.
4.3 Find the limit of

$$f(x, y) = \sin(x)/x$$

as x approaches 0.
4.4 Obtain the limits from the right and from the left for the function $sin(1/x)$ as x approaches 0.
4.5 For the function

$$f(x, y) = \frac{x^2 + y^2}{x^2 - y^2}$$

First find the limit as x approaches 1 and then as y approaches 2.
4.6 Obtain the limit of the function

$$f(x, y) = x^2 - y^2$$

First as y approaches 2 and then as x approaches 1.

Section 4.2

4.7 For the sequence

$$\{a_n\} = \frac{\log(n)}{n}$$

find the limit as n approaches infinity. The first term in the sequence is for $n=1$.
4.8 Evaluate the limit of the sequence $\{a_n\}$ for

$$a_n = \frac{(-1)^n}{n}$$

The first term in the sequence is for $n=1$.

4.9 Find the limit for the sequence

$$a_n = \frac{(-1)^n}{n^2 + 1}$$

.

Section 4.3

4.10 Find out if the function

$$f(x) = 1 - \sqrt{1 - x^2}$$

is continuous at $x=0$.

4.11 Obtain the lateral limits at $x=1$ for the function

$$f(x) = \frac{\ln(x) + \tan^{-1}(x)}{x^2 - 1}$$

and find if it is continuous.

4.12 Find out if the function

$$f(x) = \cos^{-1}\left(\frac{1 - \sqrt{x}}{1 - x}\right)$$

is continuous at $x=1$.

Section 4.4

4.13 Obtain the derivative of

$$f(x) = 3x^4 + 1$$

4.14 Obtain the derivative of

$$f(x) = -7e^{x+1} + 1$$

.

4.15 Obtain the third derivative of

$$f(x) = \cos(\sin x)$$

4.16 Obtain the first partial derivatives of the function

$$f(x, y) = \cos(x + y)$$

4.17 Obtain the second partial derivatives of

$$f(x, y) = 4 - 4x^2 - 2y^2$$

Section 4.5

4.18 Evaluate the integral

$$\int_{-0}^{5} \sqrt{1 + x^2}\, dx$$

4.19 Obtain the indefinite integral for the function

$$f(x) = \sec(x)\tan(x)$$

4.20 Obtain the indefinite integral for the function

$$f(x) = e^{2x}$$

4.21 Evaluate the double integral

$$\int_{0}^{2\pi} \int_{0}^{\pi/2} \sin(x)\cos(y)\, dx\, dy$$

4.22 Obtain the improper integral

$$\int_{0}^{\infty} e^{x^{-2}}\, dx$$

4.23 Using the instruction **quad** evaluate the integral

$$\int_{0}^{10} x^2 e^{x^{-2}}\, dx$$

4.24 Evaluate the integral

$$\int_{z=0}^{3} \int_{y=-1}^{2} \int_{0}^{1} xyz^2\, dx\, dy\, dz$$

4.25 Evaluate the line integral

$$\oint_{C} s^2\, ds$$

where C is the path from $s = 0$ to $s = 1 + i$.

Section 4.6

4.26 Evaluate the series

$$\sum_{n=0}^{\infty} \frac{x^{2n+1}}{2n+1}$$

Compare the result with the expression with the series expansion for $tan^{-1}(x)$

when it is written in terms of the natural logarithm.

4.27 Find the limit of the series

$$\sum_{n=0}^{\infty} (-1)^n (n+1) x^n$$

4.28 Evaluate the limit of the series

$$\sum_{n=0}^{\infty} \frac{n^2}{2^n}$$

4.29 Find the limit of the series

$$\sum_{n=0}^{\infty} (n+1)(x+1)^n$$

Section 4.7

4.30 Solve the differential equation

$$y''' + 3y'' - 2y' + y = 0$$

4.31 The differential equation

$$y'' - y' = \cosh(x)$$

has the initial condition $y(0)=2$ and $y'(0)=12$. Find its solution.

4.32 Find a solution for the following system of linear differential equations

$$x' = 3x - y$$

$$y' = 9x - 3y$$

with the initial conditions $x(0)=1$ and $y(0)=0$.

4.33 Solve the fourth-order differential equation:

$$y^4 + 3y^3 = x + e^x$$

4.34 Find the characteristic polynomial and its roots for the differential equation

$$y^4 + -7y^3 + 9y'' + 5y' + 6y = 0$$

4.11   References

[1] R. Larson, and B. H. Edwards, Calculus, Brooks/Cole Thomson, Belmont, CA, 2009.

[2] H. Anton, I. C. Bivens, and S. Davis, Calculus Multivariable, 9th Ed., J. Wiley and Sons, N.Y., 2009.

[3] R. L. Borrelli, C. S. Coleman, Differential Equations: A Modeling Perspective, Matlab Technology Resource Manual, 2nd Ed., J. Wiley and Sons, N.Y., 2004.

[4] W. Bober, C.-T. Tsai, and O. Masory, Numerical and Analytical Methods with MATLAB, CRC Press, Boca Raton, FL, 2009.

Chapter 5

Plotting with MATLAB

5.1 Introduction

A very useful and powerful MATLAB characteristic is that very high quality and very high complicated plots can be done very easily. Thus, two-dimensional and three-dimensional plots are possible to do. This makes data visualization an important part in MATLAB usage. In this chapter we cover the instructions to produce many of the plots available in MATLAB. Also, we see the way the properties for each plot can be changed.

The chapter is organized as follows: first we treat the several one two-dimensional plots as well as the way their characteristics can be changed. Then, we move to three-dimensional plots. The next topic we introduce is the concept of handles, which is not exclusive to plots, but it is mostly used with plots. Finally, we give a brief but detailed coverage of object hierarchy in MATLAB.

5.2 Two-Dimensional Plotting

The basic instruction to plot a function is plot (X, Y). Here X and Y are vectors with the information about the independent and the dependent variables. We generate first the vector X and then we evaluate the function we wish to plot at the points of vector X to find vector Y. With the instruction plot, MATLAB generates a new window called Figure 1 where the plot is displayed. The vector X can be generated with the instruction linspace as in

$$x = \text{linspace} (x_1, x_2, n)$$

This produces a set of points equally spaced. The first point is x_1, the last point is x_2, and the distance between points is (x_2-x_1)/(n-1). For example,

$$x = \text{linspace} (0, 5, 6)$$

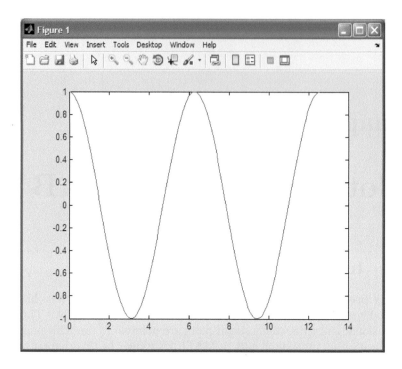

Figure 5.1: Plot of cos x.

x =
     0 1 2 3 4 5

We see that the first point is 0, the last point is 5 and there are 6 points. The distance between consecutive points is $(5\text{-}0)/(6\text{-}1)=1$. Then, the points are stored in a row vector of dimension 6. If we now wish to plot the function cos(x) from 0 to 2 with 100 points between these limits, we use

x = linspace (0, 4*pi, 200);
y = cos (x);
plot (x, y)

we obtain the plot shown in Figure 5.1. We can also use the **plot** instruction with

x = linspace (0, 2*pi, 100);
plot (x, cos(x));

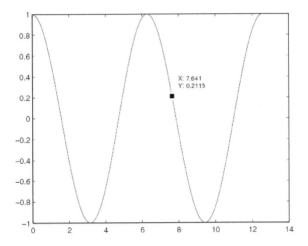

Figure 5.2: Use of the cursor.

Note that we obtain the same result as Figure 5.1 above. We also see that this window is labeled Figure 1 in the upper left corner. Also note that there is a toolbar above the plot. The toolbar has, besides the regular icons as New, Open, Save, and Print, icons for View and Cursor. The View icon will be discussed later when we talk about three-dimensional plots. The Cursor icon is used to see the coordinates of a particular point in the plot. To use the cursor we just click on the cursor icon and place the pointer on the curve and the coordinates for that point are displayed, as shown in Figure 5.2. By right-clicking we get options to delete the cursor data point or to add another data point.

If we wish to plot two functions on the same figure, we use the instruction hold on, as shown below. If we use another instruction plot, the resulting curve will appear in the same figure. The instruction hold on can be cancelled with an instruction hold off. It is also cancelled if we close the figure. For example, let us plot in the same figure the functions sin x and cos x:

```
x = linspace ( 0, 2*pi, 100);
y = cos(x); plot ( x, y);
hold on
z=sin(x)
plot ( x, z)
```

The resulting plot is shown in Figure 5.3.

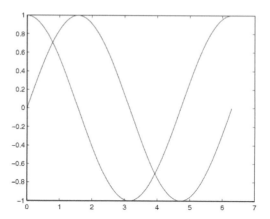

Figure 5.3: Two curves in the same plot using **hold on**.

We can also plot two functions by defining pairs x, y. For example, to plot the functions *sin 2x* and *cos 3x* in the same figure we use:

```
x = linspace (0, 2*pi,100);
y1 = sin (2*x);
y2= cos (3*x);
plot (x, y1, x, y2)
```

The results are shown in Figure 5.4 Note that each trace has a different color. We can add text information to the plot using the instructions xlabel, ylabel, title and legend. They add text to the x axis, y axis and to the plot, respectively. For our plot we can add:

```
xlabel ('x axis')
ylabel ('y axis')
title ('Functions sine and cosine');
legend ('sin (x)', 'cos(x)')
```

The figure now looks as shown in Figure 5.5.

Other plots may have a semilogarithmic axis, either for the x axis or the y axis. For a semilog x axis we use a semilogx and for a semilog y axis we use a semilogy instruction

```
x = linspace (0, 2*pi, 100);
y= sin (x);
semilogx (x, y)
```

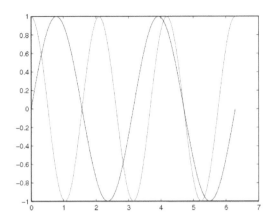

Figure 5.4: Two traces in the same figure using pairs **x**, **y**.

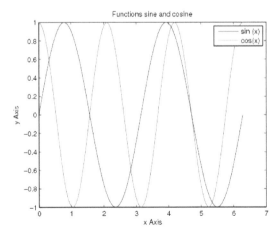

Figure 5.5: Text information added to plot of **sin x** and **cos x** functions.

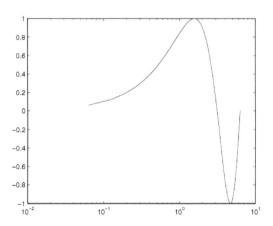

Figure 5.6: Semilog plot for **sin(x)**. The **x** axis is in a log scale.

The plot is shown in Figure 5.6. Note that the x axis is in a logarithmic scale while the y axis is linear. For a semilog y axis we use instead semilogy as in

```
x = linspace (0, 2*pi, 100);
y = exp(x.^2);
semilogy (x, y)
```

The plot is shown in Figure 5.7.

Another way to have several functions in the same plot is to form a vector of functions and plot this vector. In this case the vector of x points must be a column vector. For example, to plot the functions sin x, cos x, sin x*cos x, x*sin(x), we proceed in the following way:

```
x = linspace (0, 2*pi, 100)';
y = [sin(x), cos(x), sin(x).*cos(x), abs(x).*sin(x)];
plot(x, y)
xlabel('x-radians ')
ylabel('Four functions')
title('Multiple plot')
legend('sin x', 'cos x', 'sin x *cosx', 'abs(x)*sin x')
grid on
```

The plot is shown in Figure 5.8.

We have added an instruction grid on to generate a grid on the plot. The grid can be disabled with the instruction grid off.

The number of points in the vector x determines the plot smoothness. If the number of points in the trace is small, the plot could be a rough approx-

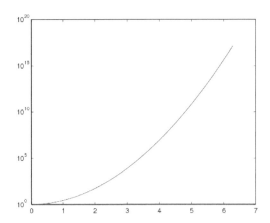

Figure 5.7: Semilog plot for **exp(x^2)**. The **y** axis is in a log scale.

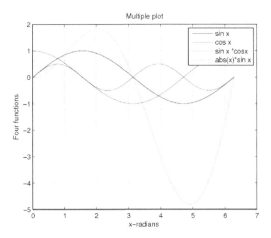

Figure 5.8: Multiple plot.

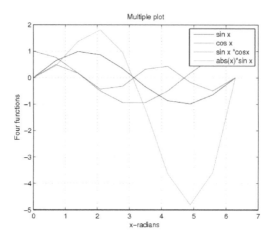

Figure 5.9: Plot with a few points for the x axis.

imation to the true trace. For example, if the number of points is very small
as in

x = linspace (0, 2*pi, 10)';

which produces a vector with only 10 points, the previous plot is now as shown
in Figure 5.9. Thus, it is better to have many points in the vector x.

5.2.1   Plotting from the Workspace

MATLAB allows to produce a plot directly from the Workspace. To see how
this is done, let us consider a new MATLAB session. Now we generate the
vectors x and y as follows:

x = 0: 0.01: 10;
y = exp(x).*sin(2*pi.*x);

We now take a look at the Workspace window and there we see icons for the
variables just created as shown in Figure 5.10. Also we see that the Workspace
window has a toolbar with some icons disabled. Now we select the variable
y which corresponds to data for the function exp(x)*sin(x) and as soon as we
select it those icons are enabled and ready to use. One of the icons is to plot
the variable selected. Figure 5.11 shows the pull down menu next to the plot
icon and it shows the different plots that can be produced. When we click on
it, the plot is produced as shown in Figure 5.12. We will cover those other
plots in the following sections.

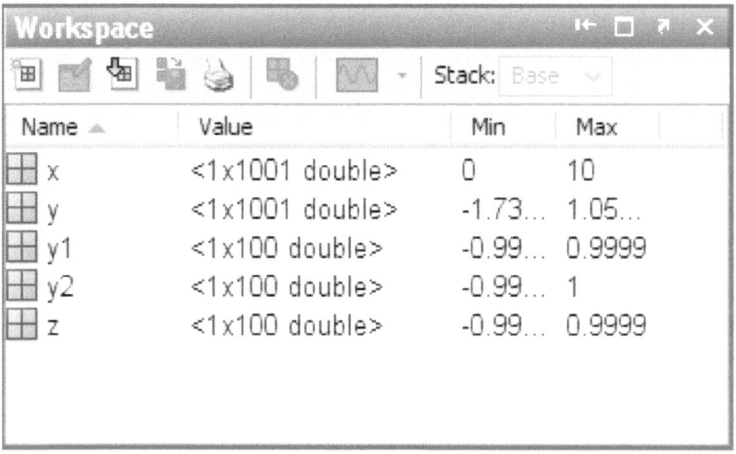

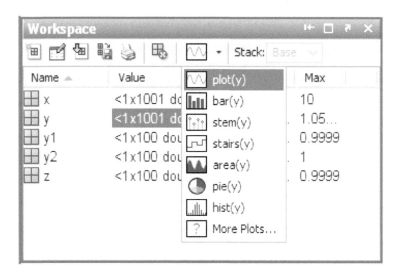

Figure 5.10: Workspace window with variables x, y created.

Figure 5.11: Selection of the plot type.

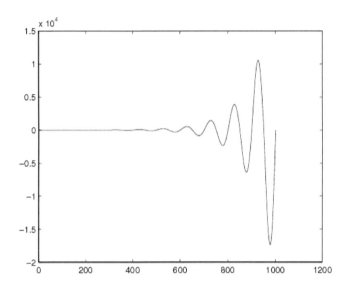

Figure 5.12: Plot for the variable y.

5.2.2   Mesh Plots

For the time being, we show how to produce a three-dimensional plot of the mesh type, covered in Section 5.5. We consider the function

$$f(x, y) = |x| + cos \, |y| \tag{5.1}$$

The necessary variables are defined by

    [x, y] = meshgrid(-10: 0.1: 10);
    f = abs(x)+cos(abs(y));

In the Workspace we select the three variables, x, y, z, (see Figure 5.13) and in the plot pull down menu we select More Plots and select 3D Surfaces and the mesh option as shown in Figure 5.14. After selecting mesh we obtain the plot shown in Figure 5.15.

5.3   Plot Options

There are several options we can use to give more information or make it more informative to users. The general form of the instruction plot is

    plot (X1, Y1, S1, X2, Y2, S2,...)

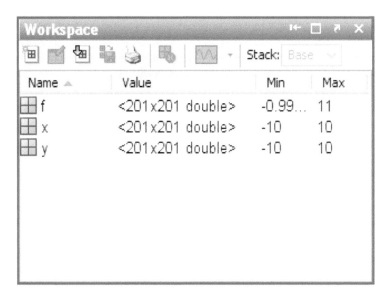

Figure 5.13: Variables selected for the mesh plot.

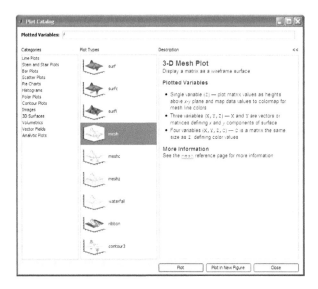

Figure 5.14: Dialog window to choose the desired plot.

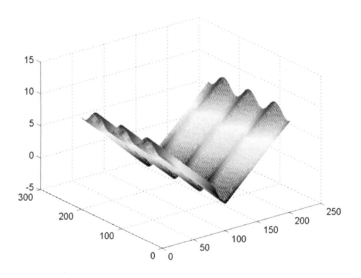

Figure 5.15: Mesh plot for the variable z.

Figure 5.16: Icon for editing the figure properties.

Where Xi, Yi have the information about the traces to plot, such as the color, width, form, etc. The terms Si are strings containing the information according to Table 5.1 and they are strings where we can specify color, markers, marker size, trace width, and trace styles. Table 5.2 shows the different options for trace color, style and markers. These options can also be changed with the property editor. To use this editor, we click on the Edit Icon, shown in Figure 5.15 and this action allows us to select the element of the figure we use to edit. We just right click and select the parameter we wish to change. In this menu we can change line width, marker, color, marker size, and line style.

When we need to draw traces, the colors are different from each one. The colors are selected by default starting with blue and continuing downward in Table 5.1. If the background is white, this color is not used for the traces. The background color can be changed with the instruction whitebg, for example:

whitebg('blue')

Table 5.1: Codes for trace color

Color code	color
b	blue
g	green
r	red
c	cyan
m	magenta
y	yellow
b	black
w	white

Table 5.2: Trace marker

Marker code	Marker
.	point
o	circle
x	cross
+	plus sign
*	asterisk
s	square
d	diamond
v	downward pointing triangle
^	upward pointing triangle
<	left pointing triangle
>	right pointing triangle
p	five-point star
h	six-point star

Table 5.3: Trace style

Style code	Trace style
-	Solid line
:	points
-.	Dash and point
- -	Dashed line

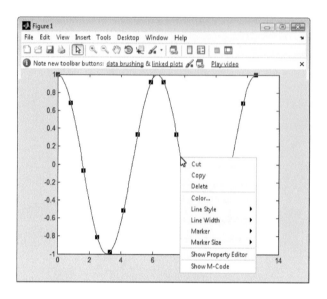

Figure 5.17: Edition of trace properties.

changes the background color from white to blue. We can go back to the previous color with the same instruction.

To place text on a plot we use the instruction text with the following format:

text (x, y, 'string')

x, y are the coordinates where we start the text on the plot and string contains the text. For example,

```
x = linspace(0, 2*pi, 100);
y1 = sin(x);
y2 = cos(x);
plot( x, y1, x, y2)
text(2, 0.5, 'This is a plot of sin and cos')
```

and we get the plot shown in Figure 5.18.

By default, MATLAB uses the information from vectors x, y to assign upper and lower bounds in the axes. However, the user can adjust them with the instruction axis that has the following format:

axis ([x_initial   x_final   y_initial   y_final])

For example, for our last plot we can use

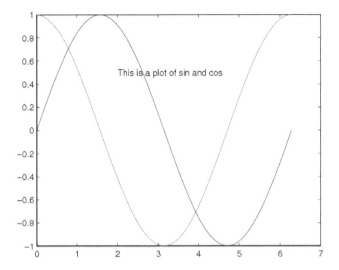

Figure 5.18: Text on a plot.

axis ([0 5 -2 2])

This instruction specifies that variable x goes from 0 to 5 and variable y goes from -2 to +2. The resulting plot is shown in Figure 5.19.

Another way to make changes in the axes' limits in a plot is through the use of the figure's menu. From the figure's Edit menu, select Edit→Axes Properties which opens the Axis menu below the figure as shown in Figure 5.20. In this menu we see two small zones. In the zone to the left we can change the color properties to the different components of the plot such as the fonts, traces, grid, background, and we are also allowed to enter the text for the title's figure. In the right zone to the right we can change the x, y, and z axes' limits, change the scale from linear to log and vice versa, add a text to the axis, as well as to change the font for axes information.

The last icon in the figure's toolbar can be used to display the menus available for the figure. The icon has the name Show Plot Tools and Dock Figure. By making a click on this icon, we obtain Figure 5.21(a). There by clicking on any of the plot components we obtain the corresponding menu. Figure 5.21(b) shows the menu for the trace of the sine function. Note that we can give a title or a text in all the menus that we open in this way.

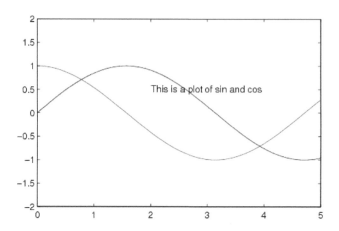

Figure 5.19: Change of axes' limits.

5.4   Other Two-Dimensional Plots

In this section we present the different two-dimensional plots available in MATLAB.

5.4.1   Polar Plot

We can make a polar plot if the function is written in polar coordinates. The instruction is

polar (θ, r, s)

where θ, r are the polar coordinates and s is a string with the same options as in the instruction **plot**. For example, if we wish to plot the function, given in polar coordinates by the equation

$$r = \theta^{1/2}$$

We use

theta = linspace (0, 8*pi, 200);
r = theta.^(1/2);
polar (theta, r)

This set of instructions produces the plot shown in Figure 5.22. We see that the function we plotted is a spiral.

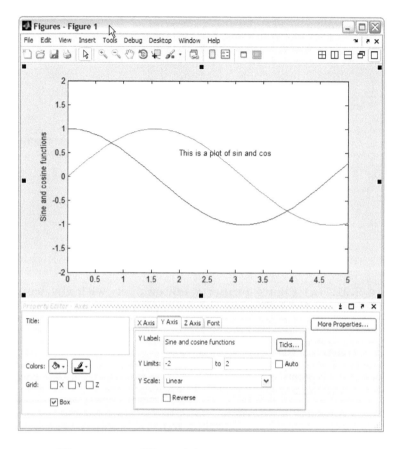

Figure 5.20: **Plot with axes menu displayed.**

5.4.2   Bar Plot

A bar plot can be obtained with the instruction

bar (x, y, s)

X, Y are the vectors with the information and S is a string for the options. Note that it is the same format used by the instruction plot. As an example, to plot 21 points from -10 to 10 we use

```
x = linspace(-10, 10, 21);
y = exp(-x.*x);
bar(x, y);
title('Bar plot')
```

The bar plot is shown in Figure 5.23.

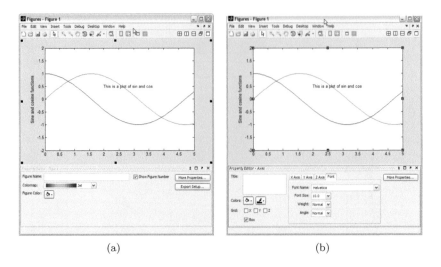

(a)   (b)

Figure 5.21: (a) Figure property editor open with the icon Show Plot Tools and Dock Figure. (b) Axes property editor.

5.4.3 Stairs Plot

The same set of points can be plotted using a **stairs** instruction. The format is

$$\text{stairs} (x, y, 's')$$

In the previous plot, by changing **bar** for **stairs** and in the title instruction we change bar for stairs we get the stairs plot shown in Figure 5.24.

5.4.4 Histogram Plot

A histogram bar is similar to a bar plot, but we only get 10 bars with a histogram instruction. The instruction is

$$\text{hist} (y, n)$$

where y is the function to plot and n is the number of bars in the plot (the maximum number of bars is 10). For example, to get a histogram for a random variable we do the following to get Figure 5.25:

```
x = linspace(-5, 5, 50);
y = randn(5000, 1);
hist(y, x);
title('Histogram plot')
```

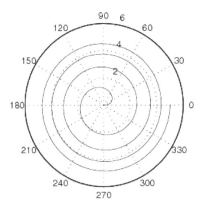

Figure 5.22: Polar plot of a spiral.

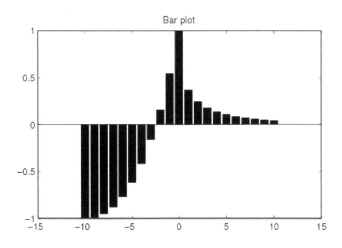

Figure 5.23: Bar plot.

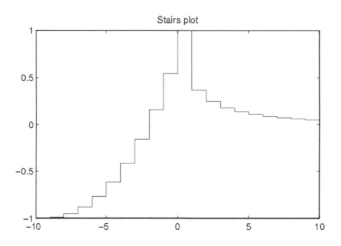

Figure 5.24: Stairs plot.

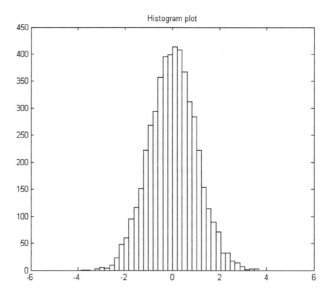

Figure 5.25: Histogram plot.

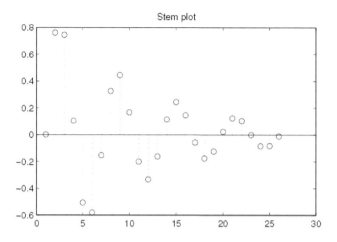

Figure 5.26: Stem plot.

5.4.5 Stem Plot

A stem plot is a plot of points. The instruction is

$$\text{stem } (x,\ y,\ 's')$$

For example, for the set of points given by exp(-t)*sin(y) we get

```
x = 0: 0.5: 4*pi;
y = exp(-0.2*x).*sin(2*x);
stem (y, ':');
```

This set of instructions produces the plot of Figure 5.26.

5.4.6 Compass Plot

This instruction plots magnitude and phase of a set of vectors. The format is.

$$\text{compass } (z) = \text{compass } (x+jy) = \text{compass } (x,\ y)$$

where z is a vector of complex numbers. For example, for vector z given by

$$z = [3+2j,\ -4+7j,\ 6-9j,\ -4-5j]$$

we get Figure 5.27 with

```
z = [3+2j, -4+7j, 6-9j, -4-5j]
compass (z)
```

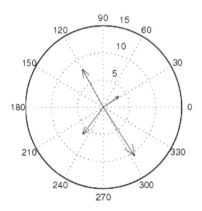

Figure 5.27: Compass plot.

5.4.7   Pie Plot

A pie plot is used to display percentages for each element in a vector or matrix. pie and pie3 generate two-dimensional and three-dimensional plots, respectively. For the vector A = [3 4 5 9 3], the pie plot is obtained with

```
A = [6 4 5 9 12];
pie(A)
```

and the result is shown in Figure 5.28.

5.5   Subplots

Sometimes we need to plot different functions in separate plots, but we need to visualize them at the same time. Fortunately, MATLAB allows us to do this by using subplots. We can generate up to four different plots in the same figure. Each plot in the figure is called a subplot. Each subplot is generated separately using the instruction subplot whose format is:

subplot (m, n, p)

This instruction divides the figure in m×n subplots arranged in a matrix form with m rows and n columns. The variable p activates each subplot. For example, for four subplots we use a 2×2 array and use subplot (2, 2, p) and p indicates the position in the array. In this case p can take values from 1 to 4. Thus, the instruction subplot(2, 2, 1) produces the plot in the (1, 1) position in the figure, subplot(2, 2, 3) produces the plot in the (2, 1) position in

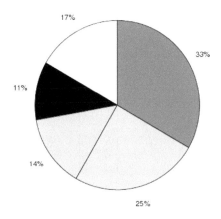

Figure 5.28: Pie plot.

the figure, and subplot(2, 2, 2) corresponds to the (1, 2) position. For example,

```
subplot(2, 2, 1)
x = linspace(0, 2*pi, 100);
y = sin(x);
plot(x, y);
title('sine function')
subplot(2, 2, 2)
y = cos(x);
plot(x, y);
title('cosine function')
subplot(2, 2, 4)
r = 5*log10(x);
polar(x, r)
title('Spiral')
```

The subplots activated are located two in the first row and one in the second row, as shown in Figure 5.29.

5.6 Three-Dimensional Plots

Three-dimensional plots, besides being more attractive to viewers, give more information to them. There are several types of three-dimensional plots, some of them are covered in this section. For a three-dimensional plot, there are three variables: x, y, z. We plot $z = f(x, y)$ which is a function of x, y. The

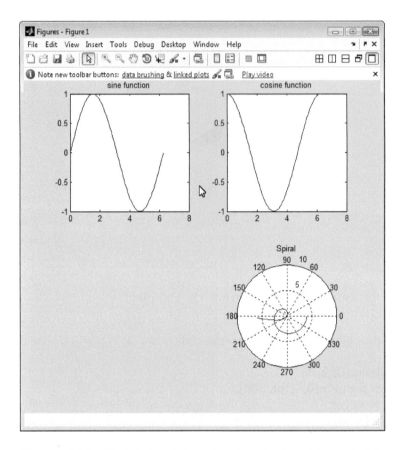

Figure 5.29: Multiple plots using the instruction **subplot**.

three dimensions are then *x, y, z*. Depending upon which type of plot we need we use the appropriate instruction.

5.6.1   The Instruction **plot3**

The instruction to plot in three dimensions is **plot3**. It has the format

$$\textbf{plot3(x, y, z, s)}$$

Here, x, y, z are the coordinates and s is a string with the options for the plot. For example, if we wish to plot the helix function, which is described by the equations

$$x = \sin t$$
$$y = \cos t$$
$$z = t$$

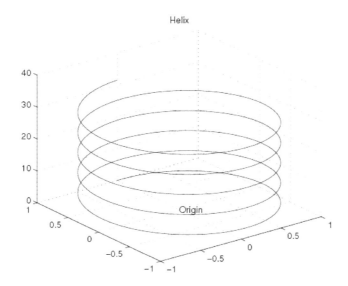

Figure 5.30: Helix plot using plot3.

we can obtain it with

```
t = linspace (0, 10*pi, 500);
plot3(sin (t), cos(t), t);
title ('Helix')
text (0, 0, 0, 'Origin')
grid on
```

The plot is shown in Figure 5.30.

Now let us suppose that we wish to plot three curves. These curves are defined by equations: $\sin x$, $\cos x$, and $\sin 2x$, in three different planes. This can be done if each function starts its values in a different value for the y variable, that is, in a different plane. The code to do this is:

```
x = linspace (0, 3*pi, 100);
Z1 = sin (x);
Z2 = cos (x);
Z3 = sin(x).*sin(x);
Y1 = zeros (size (x));
Y2 = ones (size (x));
Y3 = Y2/2;
plot3(x, Y1, Z1, x, Y2, Z2, x, Y3, Z3);
grid on
```

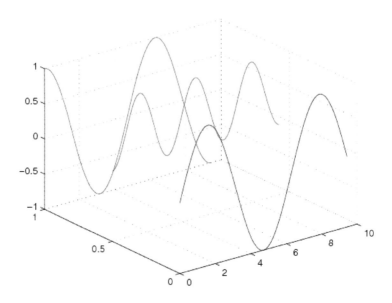

Figure 5.31: Three two-dimensional curves in a three-dimensional plot.

    title('sin x, cos x, sin x^2')

After running this code we get Figure 5.31. There we see each curve is plotted in a different plane.

5.6.2   Mesh Plot

A mesh plot is a three-dimensional surface with lines drawn forming a mesh over the surface. The functions that can be plotted are of the form

$$z = f(x, y)$$

and the instruction is

$$\text{mesh } (x, y, z)$$

Before executing the mesh instruction, we need to create a matrix for x, y values. This can be done with the instruction **meshgrid** which has the format

$$[x, y] = \text{meshgrid } (x_i, y_i: \text{ inc: } x_f, y_f)$$

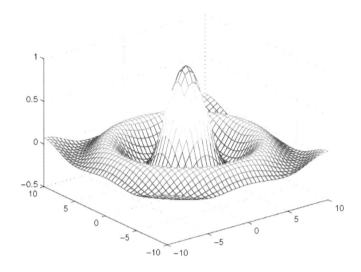

Figure 5.32: Mesh plot.

where (x_i, y_i) and (x_f, y_f) are the initial and final values for x, y, and the variable inc is the increment for x, y. For example,

```
[X, Y] = meshgrid (-10: 0.5: 10);
R = sqrt(X.^2 + Y.^2) + eps;
Z = sin(R)./R;
mesh (X, Y, Z)
```

produces the meshplot of Figure 5.32. As we can see in the screen monitor, the surface is plotted in color.

This meshplot can be made transparent by using the instruction hidden off. After executing this instruction, the new plot is shown in Figure 5.33.

The instruction mesh(x, y, z) can accept plot options, as was the case of the plot instruction. For example, for a figure in black, we get Figure 5.34 if we execute the following

```
mesh (X, Y, Z, 'Edgecolor','black')
```

Two instructions similar to mesh are meshc and meshz. meshc adds a contour map over the x,y plane and meshz a zero plane. If we change mesh for meshc and then for meshz we get the plots shown in Figures 5.35 and 5.36.

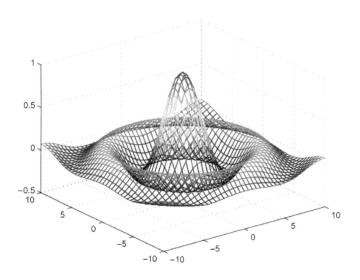

Figure 5.33: Transparent mesh plot using **hidden off**.

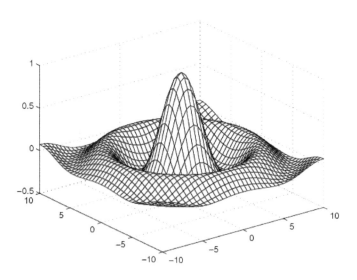

Figure 5.34: Black mesh plot using options.

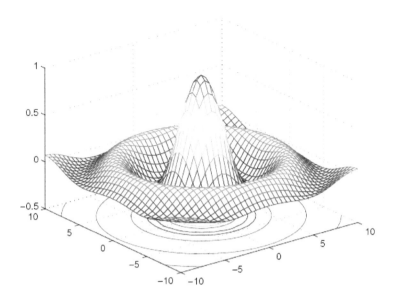

Figure 5.35: Mesh plot with a contour using **meshc**.

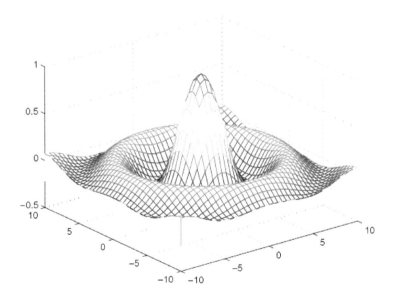

Figure 5.36: Mesh plot with a zero plane using **meshz**.

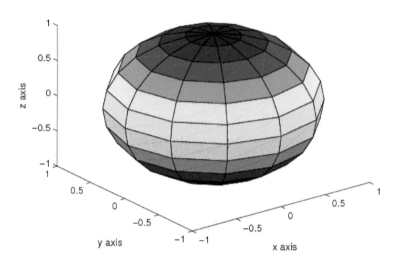

Figure 5.37: Plot of sphere with **surf**.

5.6.3   Surf Plot

The three-dimensional plot obtained with **surf** is similar to the mesh plot, except that the rectangular portions on the surface are colored. The colors are determined by the values of the vector Z and by the color map. Let us consider as an example, the surface **sphere**, available from MATLAB. If we desire to plot this surface we use the following instructions:

```
[x, y, z] = sphere(12);
surf(x, y, z)
title('Sphere plot')
grid
xlabel('x axis'), ylabel('y axis'), zlabel('z axis')
```

To obtain the plot in Figure 5.37.

In the sphere figure we can delete the black lines with the instruction

shading flat

to get Figure 5.38. If instead of **shading flat** we use the instruction

shading interp

Sphere plot

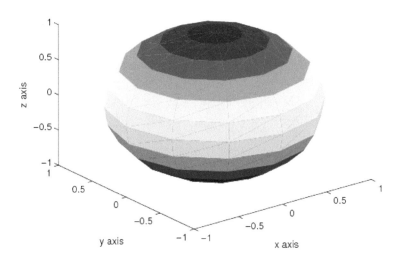

Figure 5.38: Plot of sphere without lines.

we obtain a plot with the colors smoothed, as shown in Figure 5.39.

The instruction surfl produces a plot with lighting. It has the format

surfl (x, y, z)

Using it for the sphere we produce Figure 5.40.

5.6.4 Contour Plot

The instruction Contour gives a two-dimensional plot from a three-dimensional plot. The format is

[x, y, z] = peaks (30);
contour(x, y, z, 16)
xlabel ('x axis'), ylabel ('y axis')
title ('Peaks contour')

This set of instructions produces the Figure 5.41. A variation of contour is contour3 which produces a three-dimensional contour. If in the previous example we only change contour by contour3 to obtain Figure 5.42.

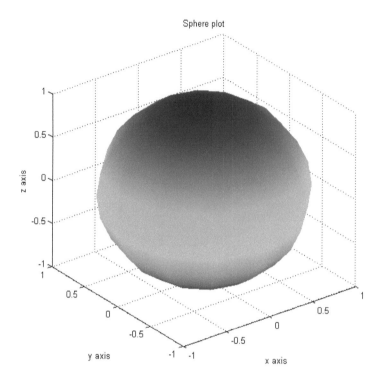

Figure 5.39: Plot of sphere with smoothed colors.

A similar contour but using pseudocolor is obtained with the instruction pcolor and the results are shown in Figure 5.43.

The instruction waterfall, also provides a contour plot but resembling a waterfall. For example, for the peaks function:

```
[x, y, z] = peaks (30);
waterfall(x, y, z)
xlabel('x axis'), ylabel('y axis'), zlabel('z axis')
```

These instructions produce the plot shown in Figure 5.44.

The instruction quiver gives directional lines to a contour plot. We have to define a differential vector DX, DY using a gradient instruction. For example,

```
[X, Y, Z,] = peaks (30);
[DX, DY] = gradient (Z, 0.5, 0.5);
quiver (X, Y, DX, DY)
```

which produces the plot in Figure 5.45.

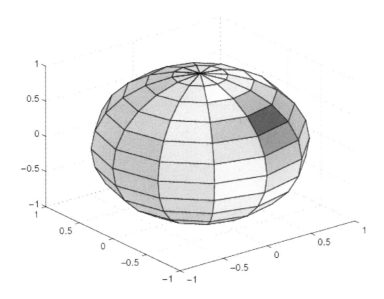

Figure 5.40: Surface plot with lighting.

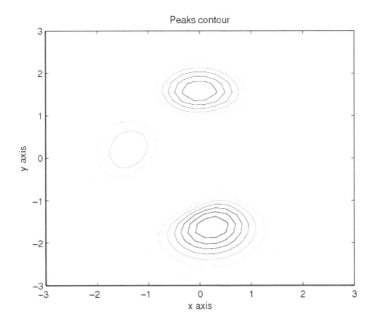

Figure 5.41: Contour plot.

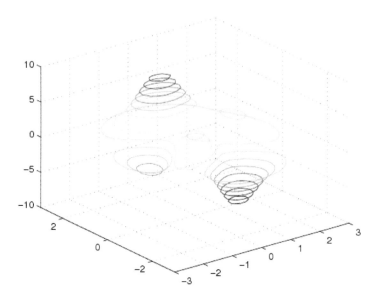

Figure 5.42: Three-dimensional contour plot.

The instruction clabel adds values to the height of a contour plot. We have first to generate a contour plot. This can be done in the following way:

```
[x, y, z] = peaks (30);
cs = contour (x, y, z, 10);
clabel (cs)
xlabel(x axis'), ylabel('y axis')
title ('Contour of peaks with values.')
```

The result is shown in Figure 5.46.

5.7   Observation Point

If we observe any of the three-dimensional plots from the previous sections, we see that the user sees them from a predetermined view point. That view point is in spheric coordinates, the distance r, the azimuth angle ϕ and the elevation angle λ as is shown in Figure 5.47.

The elevation is the angle between the radius r and the plane xy. The origin in the figure is given by (xmin, ymin, zmin). The azimuth angle is the angle formed by the projection of radius r on the xy plane and the -y axis. Default values for azimuth and elevation angles are -37.5° y 30°, respectively.

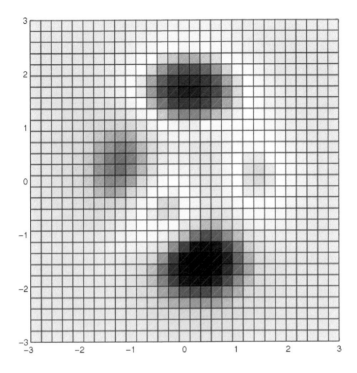

Figure 5.43: Plot with pcolor.

A two-dimensional plot has default values of 0° and 90° for the azimuth and elevation angles, respectively. The view point can be changed with the instruction view by giving the new values for azimuth and elevation angles. The format is

view(elevation, azimuth)

For example, consider the mesh plot from Figure 5.32. We can change the view point to an elevation angle of 30 and an azimuth angle of 60 with

view ([30 60])

To obtain Figure 5.48, which is the same mesh plot, but from a different view point.

Alternatively, we can use the rotation button in the figure's toolbar. This button is shown in Figure 5.49. To rotate the plot, just click once on the Rotate button and then position the cursor on the figure. Wait for a few seconds and then the azimuth and elevation angles will appear on the lower

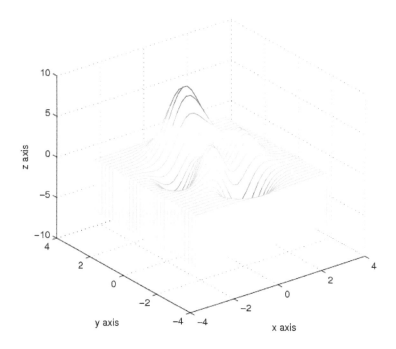

Figure 5.44: **Waterfall plot.**

left corner of the figure. Then just move the mouse around the figure with
the left-hand mouse button to move the plot.

5.8   Structure of Objects in MATLAB

If we take a closer look at the figures generated so far in the chapter, we see
that all of them are Figure 1. This is so because each time MATLAB produces
a Plot, it creates a new figure window, if none is open, or it plots the new plot
in the current figure, deleting whatever was plotted there (unless there is an
active hold on). To open several windows, each one with a different plot, we
use the instruction figure. Its format can be any one of the following:

h = figure
figure (h)

If we use the format h = figure, a new figure is created and we call h the
handle of the figure. Each time we use the instruction figure, we open a new
window. Then, the next plotting instruction will appear in that new figure.

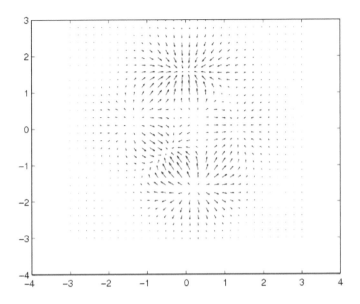

Figure 5.45: **Contour** plot with directional lines.

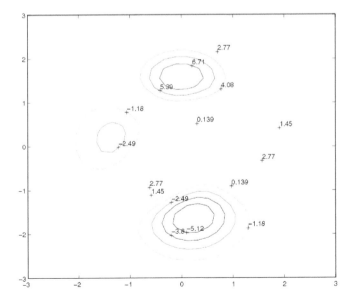

Figure 5.46: **Contour** plot with values.

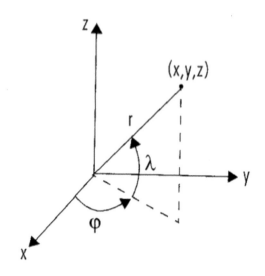

Figure 5.47: View point for a three-dimensional plot.

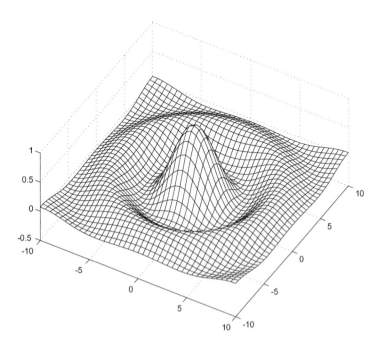

Figure 5.48: Mesh plot with a different view point.

Figure 5.49: Rotate button to change the view point.

On the other hand, the instruction figure(h) makes active the window named
Figure No. h and places it in front of all the other open windows. If Figure
No. h does not exist, it is created and it is assigned handle h. The handle
number must be an integer in order to be a figure's number. For example,

f1 = figure
    f1 =
        1

This creates a new window with handle value equal to 1 and named Figure
No. 1. Now, if we repeat the instruction as:

f2 = figure
    f2 = 2

creates another window, this time for Figure No. 2 with handle=2. Note that
as we continue this process, the handles are increasing in value consecutively.
Thus, 3 would be the next handle, and so on. We can also assign a handle to
any figure. For example, if we wish that our next window is Figure 5 we do
the following:

f3 = figure (5)
    f3 =
        5

We now have three figure windows open for Figures 1, 2 and 5. Our next step is
to draw some plots on them. We do this with the following set of instructions:

x = linspace (0, pi, 25);
y = sin(x);
plot(x, y)

We see that the plot was drawn in Figure 5 because this is the active window.
This is shown in Figure 5.50. If we now write:

figure(f1)

The window corresponding to Figure 1 would be the active one and if we plot
a function, it appears in this window. For example, for the instructions:

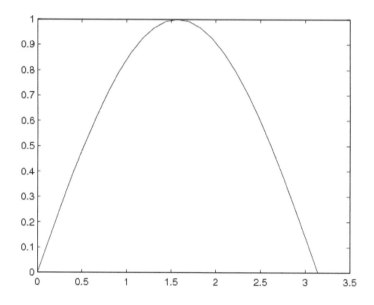

Figure 5.50: Plot in **Figure 5.**

```
y2 = cos (2*x);
plot (x, y2);
```

we see that this function is plotted in the window corresponding to Figure 1.

Other instructions useful for figure windows are: Instruction shg makes the active figure to be displayed on top of the other windows. Instruction clf deletes the active windows, while close(h) closes the window with handle h and close is used to closed the active figure. The instruction close all closes every figure window.

5.8.1 Handles for Other Objects

In every window, there are several types of objects, each one associated with a handle. For example, for the plot shown in Figure 5.48, the axes and the trace have handles. To see how handles are assigned, we write the last plot instruction as

```
handle_y = plot (x, y)
```

And what we get is

```
handle_y =
    101.0042
```

This is the value for the plot's handle. This value is 101.0042, but it may be different for different computers and/or sessions. The handle's value is not important by itself. It is more important to know the name of the variable where the handle's value is stored. Note also in this case that the handle is not an integer as is the case of figure's handles. Now let us generate a new figure and get the handle for it:

```
x = linspace (pi/2, 2*pi, 100);
y1 = sin(x);
y2 = cos(x);
y3 = sin(x) + cos (2*x);
handle_123 = plot(x, y1, x, y2, x, y3)
    handle_123 =
        127.0010
        101.0048
        128.0006
```

We see that there are three traces in the figure and each one has its own handle.

Any text on a figure has also a handle. We can find out what is the handle with:

```
handle_text_y = ylabel ('y-axis')
    handle_text_y =
        129.0006
handle_text_x = xlabel( 'x-axis')
    handle_text_x =
        130.0006
```

5.8.2 Axes Handles (gca)

To find the value of the handle of a set of axes we write **gca**. For example,

```
handle_axes = gca
    handle_ejes =
        100.0011
```

5.8.3 Object Properties

To display object properties we use the instruction **set**, as in

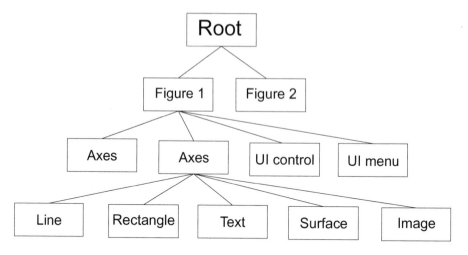

Figure 5.51: Hierarchy of MATLAB objects.

set (handle)

For example, for the last text written in the last plot we use

set (handle_ text_y)

To change the properties of an object we use the format

set (handle, Property, value)

For example, to change the text font:

set (handle text_y, "fontname", arial);
set (handle _text_y, "fontsize", 21)

5.9   Hierarchy of MATLAB Objects

Windows, graphs and every object in MATLAB have a hierarchy. They have
a hierarchic structure due to the different types of objects in each of them.
Such hierarchy has a tree-like form, as shown in Figure 5.51.

According to the figure, the most general object is the main MATLAB
window. This object is the root for all other objects as shown in Figure 5.51.
There is only a root object. The root object, the MATLAB window, can have
several windows such as the Workspace, a figure window, a properties window,
etc. Each of these windows can have objects such as axes, titles, traces, menus,
controls, etc. Finally, each of these objects can have lower hierarchy objects

such as lines, rectangles, patches, surfaces, images, and text. Some objects are called parents. The objects below a parent object are called children of that object. Thus, an object can be a parent and have children, and at the same time be a child of a parent object with higher hierarchy.

When an object is deleted from MATLAB, its children objects are automatically deleted too.

The handles of the active window, axes actives, and the active object can be obtained with the instructions

gcf (get current figure)
gca (get current axes)
gco (get current object)

Any object can be deleted with the instruction:

delete (handle)

5.10 Concluding Remarks

Plotting in MATLAB is a very simple task, even in the most complex cases. We have seen that even three dimensional and surface plots require only a few instructions to obtain them. We also covered the way plots are organized hierarchically requiring the specification of the handle to describe its hierarchy. The instructions to plot function in MATLAB have been covered and they are the most used and useful in the visualization of data.

5.11 Exercises

Section 5.1

5.1 Plot the function cosine in the interval from -2 to 2.
5.2 Plot the polynomial $p(x) = x^2 + 2x - 3$ in the rank [0, 10].
5.3 To the plot from Exercise 5.2, add the title "polynomial $p(x) = x^2 + 2x - 3$".
5.4 Plot $C(x) = x^3 - 2x$. The rank of x is from -2 to +2. Use a linear scale and a log scale. $C(x)$ is a third order Chebyshev polynomial.
5.5 In the same figure, plot: $\sin(x), x^2 + 1, x^3 - 3$, in the rank -2 to 3.14.
5.6 From the Workspace, plot $\cosh(x)$ in the rank -1 to +1.
5.7 From the Workspace, plot from -10 to 10 and in the same figure, the functions $\sin(xy)$ and $p(x) = x^2y + xy^2 - xy - +2$.

Section 5.2

5.8 Make a plot for the function $\sin(x)$ from -2 to 2 and then change the trace to red.

5.9 Using the menu for the figure, add the text "Plot for the hyperbolic sine" to a plot for the function sinh(x). Then add to the x axis the text "x-axis."

Section 5.3

5.10 Plot in polar coordinates the plot for the function sin(x) in the rank from 0 to 8.

5.11 Make a stem plot for the sequence given by x = [1, -1, 0, 3, -7, 8].

5.12 Make a bar plot for the function cos(x) where x goes from 10 to 20 and with 20 points in this interval.

5.13 Realize a compass plot for the points given by z = linspace(j, 5+2j,10).

5.14 Obtain a pie plot for the function Y = sin(x) where x =[1, 3, pi/2, 2].

Section 5.4

5.15 Plot in a subplot the following functions: $\sqrt{x}$ in the rank $x = [0,5]$ and ln(x) in the interval $x = [0.1, 10]$.

5.16 In the same figure, plot the functions sin(x) in the interval $x = [0, 5\pi]$, cosh(x) in the interval $x = [0, 10]$, and $x^2 - 3\log(x)$ in the interval $x = [0.1, 10]$.

Section 5.5

5.17 Using the instruction plot3, plot the functions $\sin(x)\cos(x)$, $g(x) = x^{1/3}$, and $h(x) = \sin(x)/x$ from $x = \pi$ to 2π.

5.18 Make a plot for $X = [0:0.01:10]$ of the functions $Y = 100\sin(X)$, $Z = \cos(X)$ using plot3. Add a grid, a title, and names for each axis.

5.19 Make a surf plot for $z = \sin(x)\sin(y)$ in the range $[x, y]$=meshgrid(0, 2 $*$ π).

5.20 Obtain a mesh plot for the function $f(x, y) = x^2 + y^2$. Choose an appropriate range for the independent variables. Repeat for the function $g(x, y) = x^2 - y^2$.

5.21 Use meshc and meshz to plot the function $f(x, y) = |xy|$.

5.22 Plot $z = \sin(|xy|+|xy|)$ using the instruction surf.

5.23 Make a surf plot for $z = (x - y)^2$ and apply the instructions shading flat, shading interp, surfl.

5.24 Obtain a plot of $z = \sqrt{x^2 + y^2}$ using surfl.

5.25 Use the instruction contour3 to make a plot for $z = \sqrt{|xy|}$ in the grid $[x, y] = (-1, 1)$.

5.26 Make a contour plot for the function $f(x, y) = 1/(1 + x^2 + y^2)$.

Section 5.6

5.27 Change the observation point for the plot obtained in Exercise 5.17 to [10,40].

5.28 Using the icon for **Rotate** 3D and change the observation point to the plot from Exercise 5.19 so we look at it from the z-axis.

5.29 Plot in two different windows the functions from Exercise 5.15.

5.30 Plot in three different figures the functions from Exercise 5.16. Make the figures to be numbers 19, 51 and 57.

5.12 References

[1] P. Marchand and O. T. Holland, Graphics and GUIs with MATLAB, 3rd Ed., Chapman and Hall/CRC, Boca Raton, FL, 2003.

[2] D. C. Hanselman, Mastering MATLAB 7, Prentice Hall, Inc., Piscataway, NJ, 2004.

[3] MATLAB Graphics, The MathWorks, Inc., Natick, MA, 2009.

[4] M. E. Herniter, Programming in MATLAB, Brooks/Cole Thomson, Pacific Grove, CA, 2001.

[5] H. Moore, MATLAB for Engineers, 2nd Edition, Prentice Hall, Inc., Piscataway, NJ, 2008.

[6] S. Nakamura, Numerical Analysis and Graphic Visualization with MATLAB, Prentice Hall, Inc., Piscataway, NJ, 1995.

Chapter 6

Programming in MATLAB

MATLAB provides a very powerful high-level programming language, embedded in a computational environment, known as m-language. A very important advantage, when compared to other programming languages such as C, FORTRAN, and VisualBasic, to name a few, is that variables do not have to be defined at the beginning of the program, but rather they are defined automatically as they are written for the first time in the program. Another very important advantage is that all the instructions used in MATLAB can be used in the program. Thus, instructions such as the ones used for differentiation, integration, matrix inversion, solve differential equations, etc., can be used in the program. This makes the m-language much more powerful than any other programming language. In this chapter we cover many of the instructions used to write programs using m-language and give some examples.

6.1 Creating m-files

Files containing programs using m-language are called m-files. They have an extension .m. There are two types of m-files: scripts and functions. Scripts are programs that neither have input data nor output data. On the other hand, functions do have input data and produce output data.

The variables of a script or a function are called local variables to the program. Thus, in other programs, they are not available to other programs unless they are declared as global variables.

To see the contents of a program we just need to enter the instruction type followed by the file name. For example, if there is a file containing the program quadratic.m, we do the following in the Command Window:

type quadratic

An m-file containing a function must have the following parts:

165

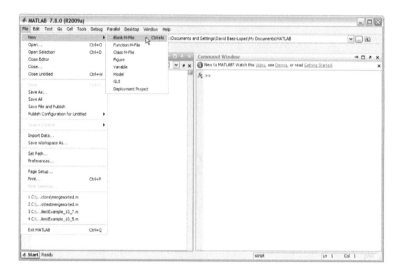

Figure 6.1: Menu to create an m-file.

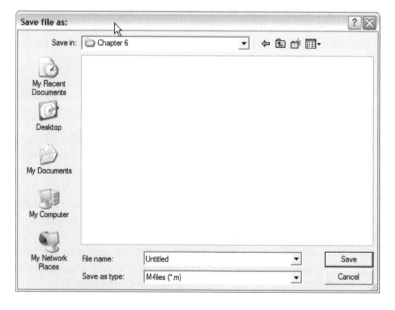

Figure 6.2: Directory to save an m-file.

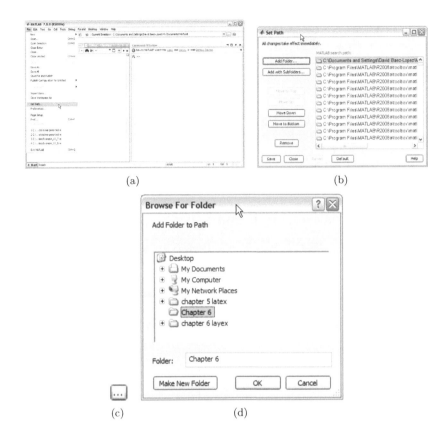

(a)   (b)

(c)   (d)

Figure 6.3: Menu to indicate the path where an m-file is saved.
(a) Choosing the Set Path, (b) With Add Folder we indicate where
we wish to save the m-file, (c) Icon for Browse Folder, (d) Browse
for Folder window.

1.- Comments that start with the symbol %
2.- Input data instructions
3.- Instructions that execute the action for the function
4.- Output instructions
5.- It may have an end instruction to end the function

To create and edit an m-file we use the M-File Editor/Debugger that comes
with MATLAB. It is shown in Figure 6.1. Once we finish writing the m-file
we save it in a directory, which in our case is called MATLAB_Chapter_6 as
shown in Figure 6.2. If we wish to save it in a different folder we indicate this
to MATLAB with Set Path in the File menu as shown in Figure 6.3(a) and in
the dialog windows we indicate in which directory search for the m-files, as

shown in Figure 6.3(b). Alternatively, we can use the Browse Folder icon in the tools bar shown in Figure 6.3(c). This opens the window shown in Figure 6.3(d) and there we select the folder where we are going to save the files we write.

6.2 MATLAB Basic Programming Instructions

In this section we present the basic programming instructions that we may require in a MATLAB m-file. We show its use with simple, but useful programs. Thus, we can appreciate the powerful m-language.

6.2.1 The Instruction if-end

A very important characteristic in any programming language is the ability to change the order in which a set of instructions is executed. This usually depends upon a set of conditions. If they are satisfied, the sequence has a certain order. If they are not, then the sequence changes the order in which instructions are executed. The set of conditions may depend upon variables in the program and/or input variables. MATLAB has instructions that do this kind of decisions. One of such instructions is the if statement. It has the general form

> **if condition**
> > **statements**
> **end**

Note that this if statement is similar to that in other programming languages.

The condition may contain logic operators. Those are shown in Table 6.1. If the condition is satisfied, then all the statements between the if and the end keywords are executed. Conditions are of the form

$$e_1 \mathscr{R} e_2 \tag{6.1}$$

where e_1 and e_2 are arithmetic expressions and $\mathscr{R}$ is one of the relation operators in Table 6.1. A small example will show the use of the instruction if.

Example 6.1 Use of the instruction if

In a school a passing grade has to be greater than 7.5. The m-file reads the input data and indicates if the student passes or fails the course. This file is created with the MATLAB editor and after writing it and before executing it we save it in the directory Chapter_6 with the name six_1.mThe m-file editor page is shown in Figure 6.4.

% File six_1.m

Table 6.1: Relation operators

Relation operator $\mathcal{R}$	Definition
>	Greater than
>=	Greater than or equal to
<	Less than
<=	Less than or equal to a
==	Equal to
˜=	Not equal to

Table 6.2: MATLAB logic operators

Operator	Description	Example Description
&	AND	(a>b)&(x<5)
|	OR	(a>b) | (x<5)
~	NOT	~ (a>b)
XOR	Exclusive OR	xor(c1,c2)

```
% Reads in a grade and writes out
% if the student passes or fails.
%
% Reads in the grade
calif = input ('Give me the grade:\n');
if calif >= 7.5
    fprintf ('Congratulations, you passed.')
end
if calif < 7.5
    fprintf('I am sorry. You failed.')
end
```

Now we run the program with a grade of 5 and a grade of 9,

```
six_1
Give me the grade:
5
I am sorry. You failed.
six_1
Give me the grade
9
Congratulations, you passed.
```

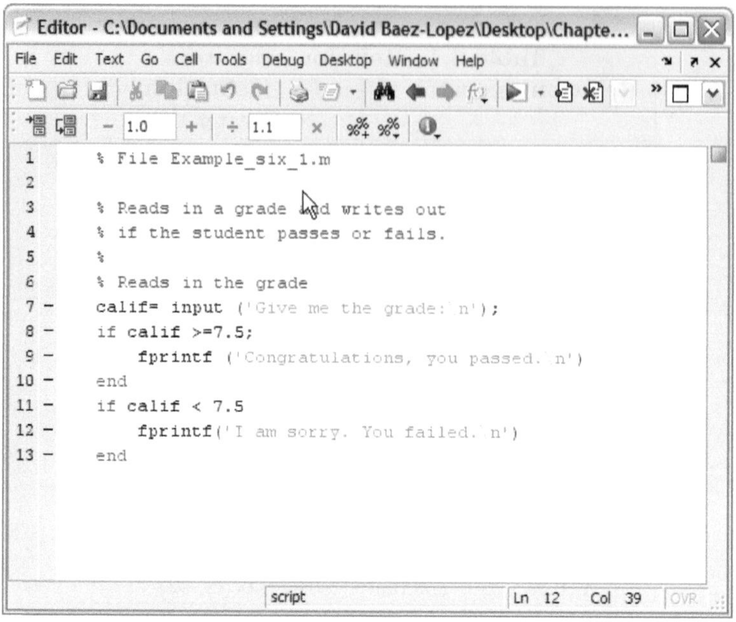

```
┌─ Editor - C:\Documents and Settings\David Baez-Lopez\Desktop\Chapte... _ □ ☒ ┐
│ File   Edit   Text   Go   Cell   Tools   Debug   Desktop   Window   Help        ⤳ ⊼ ✕ │
│  ┌─┐ ┌─┐ ┌─┐ │ ✂ ┡ ┡ ┆ ⭤ │ ⛁ ┉ ▾ │ ▶ ⬅ ⮕ ƒₗ │ ▶ ▾ ┇ ┇   »  □ ▾ │
│  ┤⬡ ┡⬡  ─ 1.0  +  ÷ 1.1   ✕  %% %% ⓞ                                  │
│  1        % File Example_six_1.m                                         │
│  2                                                                       │
│  3        % Reads in a grade ▹ᴺand writes out                           │
│  4        % if the student passes or fails.                             │
│  5        %                                                              │
│  6        % Reads in the grade                                          │
│  7 ─      calif= input ('Give me the grade: n');                        │
│  8 ─      if calif >=7.5;                                               │
│  9 ─          fprintf ('Congratulations, you passed. n')               │
│ 10 ─      end                                                           │
│ 11 ─      if calif < 7.5                                                │
│ 12 ─          fprintf('I am sorry. You failed. n')                     │
│ 13 ─      end                                                           │
│                                                                         │
│                                                                         │
│                         │ script │        │ Ln  12 │ Col  39 │ OVR │    │
└─────────────────────────────────────────────────────────────────────────┘
```

Figure 6.4: Editor page for Example 6.1.

6.2.2   The Statement **if-else-end**

A second format for the statement if is to add a statement **else**. This new statement allows programmers the use of a single if statement to execute two sets of statements. The new if statement is known as an **if-else-end** statement. Its format is

> **if condition**
>   **set of statements a1**
>   ⋮
>   **set of statements b1**
> **end**

If condition is valid then the set of statements a1 is executed. After doing this, the program continues with the instructions following the end keyword. If the condition is not valid, then the set of instructions b1 is executed. When the last one of the instructions in set b1 is executed, the program continues with the first instruction after the end keyword.

We can readily see that the statement if-else-end is equivalent to two if-end statements. In any case, the programmer has the freedom to use the instruction he/she thinks is more convenient.

Example 6.2 Use of the instruction if-else-end

We rewrite Example 6.2 using the statement if-else-end,

```
% File six_2.m
% Reads in a grade and writes out
% if the student passes or fails
% Reads in the grade
calif = input ('Enter the grade:\n');
if calif >= 7.5
    fprintf ('Congratulations. You passed.\n')
else
    fprintf ('I am sorry. You failed.\n.')
end
```

The program behaves in a similar way to Example 6.1.

6.2.3   The Instruction elseif

A third form of the statement if uses the keyword elseif. With it we can check for a greater number of conditions. The format is

```
if condition 1
    statements A1
elseif condition 2
    statements A2
elseif condition 2
    statements A2
        ⋮
elseif condition n
    statements An
end
```

The statement if-elseif-end checks which condition is first satisfied and executes the statements that follow that condition. After that, it continues with the instructions following the end keyword.

Example 6.3 Use of the statement if-elseif-end

We now modify Example 6.1 to include a statement if-else-end. This instruction can be used in the previous script to display an error if the user enters a letter instead of a numeric grade. We can correct that with the following program:

```
% File six_3.m
% % Reads in a grade and writes out
% if the student passes or fails
%
% Asks for the grade
calif = input ('Enter the grade:');
if calif >= 7.5
        fprintf ('Congratulations.  You passed.')
    elseif calif < 7.5
        fprintf ('I am sorry.  You failed..')
    else
        fprintf ('Error-You must enter a numeric grade')
end
```

6.2.4   The Statement **switch-case**

The statement switch-case can be used when we wish to check if an expression is equal to some value. It cannot be used to check conditions such as $a > 5$ or $b < 3$. However, it finds a great deal of applications where it is required to execute a set of instructions if a condition is equal to a given value. The format is

```
switch expression
case value 1
    Statements a1
case value 2
    Statements b1
case value 3
    Statements c1
otherwise
    Statements d1
end
```

value can be a numeric quantity or a string. Furthermore, each case may have one or more values, like in

```
case value1, value2,..., valuek
    statement 1
    ⋮
    statement m
```

In this case, if the expression is equal to any of the values value1, value2,..., valuek, the statements 1 to m are executed.

Example 6.4 Use of the instruction switch-case

As an example, we can modify the program from Example 6.1,

```
% File six_4.m
%
% Reads in data and writes out the result
% indicating if the student passes or fails.
%
% Enter the grade
calif = input ('Enter the grade:\n');
switch calif
case {7.5, 8, 8.5, 9, 9.5, 10}
fprintf ('Congratulations. You passed.\n')
case {0, 0.5, 1, 1.5, 2, 2.5, 3, 3.5, 4, 4.5, 5}
fprintf ('I am sorry. You failed.\n')
otherwise
fprintf ('It was not enough. You failed.\n')
end
```

6.2.5 The Statement **for**

The statement for is used to form instruction loops that have to be repeated a number of times. The format for this statement is

```
for variable = expression
    statement 1
    statement 2
       ⋮
    statement n
end
```

If variable=expression holds, then statements 1 to n are executed. The next statement to be executed is the one following the end keyword. If variable=expression does not hold, then the next instructions after the end keyword is executed. It is possible that statements 1 to n are never executed.

Note that a for statement can be used instead of an if statement.

Example 6.5 Use of the statement for

We wish to add the first 10 integer numbers. The following m-file does this task:

```
% File six_5
%
% sum = 0;
for i = 1:10
    sum = sum+i;
end
fprintf ('The result is % g.' , sum)
```

But if we only wish to add even integers between 1 and 10:

```
% File six_5a
sum = 0;
for i = 0: 2: 10
    sum = sum+i;
end
fprintf ('The result is% g' \n, sum)
```

Note that 1:10 means that variable i varies from 1 to 10 in increments of 1 in each iteration. This can also be written as 1 : 1 : 10. For the second program, to select the even integers we use 0 : 2 :10 which indicates that variable i changes value in increments of 2 starting with 0 and going up to 10. If required, the increment can be negative as in 4 : - 1 : 0 which means that it starts at 4 and it is decremented by 1 and ends at 0. Finally, the decrement can be a fraction, as in 4: 0.25 : 5. This would give 4, 4.25, 4.50, 4.75 and 5.

Example 6.6 Factorial function

The factorial of an integer n is defined as

$$n! = 1 \cdot 2 \cdots n \tag{6.2}$$

We can use a for loop to find the factorial. This can be done with the following m-file:

```
% File six_6.m
% Factorial of a non negative integer.
%
n = input ('Enter a non-negative integer: ');
n_factorial = 1;
for i = 1: n
    n_factorial_n = n_factorial *i;
end
fprintf('The factorial of % g is % g.', n, n_factorial)
```

Because MATLAB has an internal function to evaluate the factorial of a non-negative integer named factorial, we have named our variable n_factorial.

In general, users MUST NOT use names of MATLAB functions as variables or function names.

We can nest for loops. In fact, we may nest as many for loops as we wish.

Example 6.7 Use of nested for loops

We wish to add all the elements a_{ij} in an $n \times m$ matrix. This can be done with the sum

$$sum = \sum_{i=1}^{n} \sum_{j=1}^{m} a_{ij} \qquad (6.3)$$

This equation can be implemented with the following file:

```
% File six_7.m
% This m-file evaluates the sum of
% the elements of a matrix n x m.
n = input ('Enter the number of rows\n');
m = input ('Enter the number of columns\n');
% Reads in the elements
% Initialize the sum.
sum = 0;
% Reads in the matrix elements and adds them.
for i = 1: n
    % Reads in the elements of row i and adds them.
    for j = 1: m
        fprintf('Enter the matrix element %g,%g', i, j);
        a(i,j) = input(' \n');
        sum = sum+a(i, j);
    end
end
fprintf('The total sum is %g\n', sum)
```

6.2.6   The while Loop

The while loop is used to repeat a set of instructions for an unknown number of times. The difference between a while loop and a for loop is that the for loop has to be repeated a known number of times. The format for the while loop is

```
while condition
    statement 1
        ⋮
    statement n
end
```

A while loop works as follows: If condition does not hold, the instructions
after the end keyword are executed. If condition holds, then statements 1 to
n are executed. At this point, the condition is tested again. If it still holds,
then the process is repeated again and statements 1 to n are executed. If at
any time the condition is tested and it does not hold, then the next executed
instruction is the first one after the end keyword.

Example 6.8 Use of the while loop

We wish to evaluate the volume of spheres with radii 1 to 5 in steps of 1. We
can do this with the following program:

```
% File six_8.m
%
% This program evaluates the volume of spheres with radii 1 to 5.
%
% r is the radius of the sphere.
r = 0;
while r<5
    r = r+1;
    vol = (4/3)* pi*r^3;
    fprintf ('The radius is% g and the volume is %g.\n', r, vol)
end
```

When we run this file six_8 we obtain,

```
six_8
The radius is 1 and the volume is 4.18879.
The radius is 2 and the volume is 33.5103.
The radius is 3 and the volume is 113.097.
The radius is 4 and the volume is 268.083.
The radius is 5 and the volume is 523.599.
```

6.3   Functions

A function in MATLAB is a subprogram that can be used by a larger program. It is usually used to perform repetitive tasks. The format of a function
is:

```
function y = operation(x)
statements of the function
y = operation
```

Here, x is the argument or input parameter, y is the output parameter, and operation is the name of the function. Note that operation is the last instruction in the function definition.

A function can be used by another function. It can be used by itself in a recursive fashion.

As an example, let us use a function to evaluate the factorial of an integer number. We include checks in the case that the number is not an integer or it is a negative integer.

Example 6.9 Factorial evaluation using a function

We wish to evaluate the factorial of an integer. In case that the number is not a positive integer, write out a message indicating this case. The function to evaluate the factorial is:

```
function x = factorial1 (n)
factorial1 = 1;
if n == 0
   factorial1 = 1;
   else
   for i = 1: n
       factorial1 = factorial1 *i;
   end
end
x = factorial1;
```

If we execute this function for n= 3, 4 and 5 we obtain

```
factorial1(3)
ans =
6
factorial1(4)
ans =
24
factorial1(5)
ans =
120
```

Note that we did not check for n to be a non-negative integer.

Example 6.10 Use of a function inside another function

We now wish to write an m-file that evaluates the factorial checking if the number n is an integer and if it is negative.

```
% File six_10.m
% Evaluation of the factorial of an integer.
% It includes messages in the case the number given
% is a non-negative integer.
n = input ('Enter an integer number n: \n');
error =0;
if floor (n) ~= n % checks if n is an integer number.
    % floor (n) evaluates the integer part of n.
    error = 1; % n is not an integer number.
end
if n<0 % checks if n is a negative number.
    error = 2; % n is a negative number.
end
if error ==1
    fprintf ('The number entered is not an integer one.\n')
  elseif error ==2
    fprintf ('The number entered is a negative one. \n')
  elseif error ~= 1 & error~= 2
    x = factorial1 (n);
    fprintf('The factorial of %g is %g.\n', n, x )
end
```

If we run this file we obtain the following:

```
six_10
Enter an integer number n
-6
The number entered is a negative one.
```

```
six_10
Enter an integer number n
4
24
```

Thus, the factorial of 4 is 24. We run again the m-file

```
six_10
Enter an integer number n
1.4
The number entered is not an integer one.
```

Example 6.11 Solution of a second order equation

The solutions of the quadratic equation

$$ax^2 + bx + c = 0 \tag{6.4}$$

are given by

$$x_{1,2} = \frac{-b \pm \sqrt{b^2 - 4ac}}{2a} \tag{6.5}$$

A function that calculates these solutions is the function quadratic (a,b,c) saved in the file quadratic.m:

```
function [x1, x2] = quadratic (a, b, c)
% This function evaluates the roots x1, x2
% of the quadratic equation ax^2+bx+c=0
% The file is quadratic.m
discriminant =(b^2-4*a*c );
x1 = (-b+ sqrt (discriminant))/(2*a)
x2 = (-b-sqrt (discriminant))/(2*a);
```

When we run this function we get the solutions for the equation $x^2+2x+3 = 0$

```
[x1, x2]= quadratic (1, 2, 3)
x1 =
-1.0000 + 1.4142i
x2 =
-1.0000 - 1.4142i
```

And for equation $4x^2 + 8x - 60 = 0$

```
[x1, x2] = quadratic (4, 8, -60)
x1 =
3
x2 =
-5
```

6.4   Variables of Functions

The variables of the main program are independent of the variables of a function, even though they may have the same name. In general, they have different numeric values, unless they are declared as function arguments. To see this, let us consider the function test in the following example.

Example 6.12 Variables in a function

In this example we check the way variables are passed between subroutines and functions. Variables are not declared as input/output arguments.

```
function y = Test (x)
% Test function to watch variables.
fprintf ('Just entered Test. \n')
a = 1;
b = 10;
c = 100;
y = 3;
fprintf('I am inside the function Test. \n')
fprintf ('Variables are a= %g \n  b= %g \n c= %g \n' , a , b ,c )
fprintf ('Exiting the function Test. \n')
```

If this function is called from the program six_12

```
% File six_12.m
a = 4;
b = 2;
c = 0.47;
fprintf('Variables before Test: \n a=%g\n b=%g\n c=%g\n', a, b,c)
p = Test(1);
fprintf ('I am back in the main program \n');
fprintf ('Variables are \n a= %g \n b= %g \n c= %g \n' , a, b, c)
```

When we run this program we obtain:

```
six_12
Variables before Test are:
a= 4
b= 2
c= 0.47
Just entered Test.
I am inside the function Test.
Variables are
a= 1
b= 10
c= 100
Exiting the function Test.
I am back in the main program
Variables are
a= 4
b= 2
c= 0.47
```

From this run we see that the values of a, b and c given before entering the function Test are not passed to the function. In addition, the variable values for a, b and c defined inside Test are not passed to the main program. Thus, in case we need to pass them either to the function or to the main program we need to declare them as function arguments as in the following example:

Example 6.13 Passing variables as function arguments

We now show how we can pass variables from main program to functions and vice versa, as function arguments. Let us consider the following function where we have added as arguments the variables a, b, and c.

```
function [a, b, c] = Test_13(a, b, c)
fprintf ('Variables are n\ a= %g n\b= %g n\ c= %g n\' , a, b, c)
a = a*2;
b = b*10;
c = c*100;
y = [a, b, c];
fprintf('I am inside Test.')
fprintf('Variables are now \n a= %g\n b= %g \n c= %g \n', a, b, c)
fprintf ('Exiting the function. ')
```

And we use it in the main program six_13:

```
% File six_13.m
a = 4;
b = 2;
c = 0.47;
fprintf ('Values before Test \n a=%g \n b=%g \n c=%g \n' ,a, b, c)
[a, b, c] = Test_13( a, b, c);
fprintf ('Back in the main program\n');
fprintf ('Variables are \n a= %g \n b= %g \n c= %g \n' , a, b, c)
```

When this program six_13.m is run, we get the following results:

```
six_13
Variable values before Test are
a = 4
b = 2
c = 0.47
Variables in Test are:
a = 4
b = 2
c = 0.47
I am inside Test.
```

Variables have been changed to
a = 8
b = 20
c = 47
Exiting the function.
Back in the main program
Variables are
a = 8
b = 20
c = 47

We note the following: At the beginning of the program the values for the constants are a = 4, b = 2, and c = 0.47. These values are passed to the function Test_13 and they are displayed first after the text Variables in Test are:. Then they are changed and displayed from within Test as a = 8, b = 20, and c = 47. Then we exit the function Test and return to the main program where the variable values are displayed. We see that the values now are the same that were changed in the function Test.

This way to pass variables is called *call by value*. We may pass them to a function but let them unchanged in the main program. We leave this as an exercise at the end of the chapter.

6.4.1   Global Variables

In the previous section we saw how we can pass variables between a main program and a function. We learned that values do not pass just by giving the same name to variables. They need to be passed by value in order to be used in another function or in a main program. Variables defined in this way are called local variables. This may be convenient sometimes when we wish to use the same name for different variables, and we do not wish to change the variables' values. When we define a variable, MATLAB assigns it as a local variable. Unless the programmer wishes to use in another function with the same value, we have to remove the local variable restriction and declare it as a global variable. In this way, every time we refer to a given variable, declared as a global, it is going to have the same value, and if this value is changed, it is going to change in any other function and main program where the variable is declared as global. This means that if a function does not declare it as a global variable, it will not have the global variable value in that function.

The instruction to declare a variable as global is

global a, b

This means that variables a and b are global variables.

Example 6.14 Program with a global variable

To see how global variables work, let us consider the variable a = 3. We wish to have this variable in a function and in a main program. The following file shows this:

```
% File six_14.m
% Program to check local and global variables.
global a
a = 3;
fprintf('Value of variable a before function change1 a = %g \n', a)
change1
fprintf('Value of variable a after function change1 a = %g \n', a)
change2(a)
fprintf ('Value of a after exiting change2 a = %g \n', a)

function x = change1(a)
global a
fprintf ('Value of variable a after entering change1 a = %g. \n', a)
a = 7;
fprintf ('Value of a modified in change1 a = %g . \n',a)

function x = change2 (a)
fprintf ('Value of a after entering change2 a = %g . \n', a)
a = 12;

fprintf ('Valor de a modificado en prueba 2 a = %g. \n', a)
```

Now we run the main program six_14,

```
six_14
Value of variable a before entering function change1 a = 3.
Value of variable a before after entering function change1 a = 3.
Value of a modified in change1 a = 7.
Value of variable a after returning from function change1 a = 7.
Value of a after entering change2 a = 7.
Value of a before exiting change2 a = 12.
Value of a after returning from change2 a = 7.
```

As we see in the main program and in the functions, the variable a is global in the main program and in function change1 and it is only local in change2. When we run the program, the value of a is passed to a in the instruction global a. There it keeps the same value until we change it to a=7. It passes this value to the main program using again the instruction global a. The value of a is passed as an argument in the function change2. There we change its value to a $= 12$. When we exit this function, the value of a is not changed in the main program because it is a local variable in change2.

6.4.2   The Instruction return

In general, a function ends with the last instruction. But sometimes we need to end the function before the last instruction and return to the main program or function that is called the function. We can do this with the instruction return. When the sequence of instructions finds a return, it ends the function and it exits it, returning to the function or program that called it. We present an example to show how the instruction return works.

Example 6.15 Use of return in a function

Let us consider the script six_15.m which uses the function greater_smaller to find out if a number is greater than or smaller than 0.

```
% File six_15.m
x = input ('Enter the value of x: \n');
greater_smaller (x);
fprintf (' The run ends. \n')

function greater_smaller (x)
if x<0
    fprintf (' x is less than 0. \n')
    return
elseif x>0
    fprintf (' x is greater than 0. \n')
return
else
fprintf (' x is equal to 0. \n')
end
```

Now we have several runs to show how the instruction return works.

```
six_15
Enter the value for x:
0
x is equal to 0.
Run ends.


six_15
Enter the value of x:
3
```

x is greater than 0.
The run ends.

six_15
Enter the value of x:
-3
x is less than 0.
The run ends.

We observe the following: When x is greater than or smaller than 0, the function finds a **return** instruction and interrupts it before executing the remaining instructions. The control is passed to the main program which prints **The run ends**. If x=0 the function ends the if statement and then goes to the main program. In this last case the function is executed to the last instruction.

6.4.3 The Instructions **nargin** and **nargout**

The instructions **nargin** and **nargout** are used to find out the number of input and output arguments in a function, respectively. For example, for the function in Example 6.15, we have **nargin** =1, **nargout**=1. **nargin** is the acronym for Number of ARGuments in the INPUT and **nargout** for Number of ARGuments in the OUTput. **nargin** and **nargout** are variables.

6.4.4 Recursive Functions

A recursive function is a function that calls itself. This property is available for MATLAB functions. We show an example to show the recursivity in MATLAB functions.

Example 6.16 Recursive evaluation of the factorial function

The simplest example of a recursive function is the factorial function. As we already know, the factorial of a non-negative integer is defined as

$$n! = 1 \cdot 2 \cdots n \tag{6.6}$$

This can be rewritten as

$$n! = n \times (n - 1)! \tag{6.7}$$

In Example 6.15 we saw that the factorial is evaluated by the function **factorial1**. Using recursion the function can be written as

```
function x = fact_rec(n)
if n ~= 1
    x = n*fact_rec(n-1);
else
```

```
    x = 1;
end
```

When we run this program we get

```
fact_rec(6)
ans =
720
```

We see that the function is calling itself and producing the expected result.

6.5   File Management

Up to this point, input data has been entered through the keyboard. The keyboard is thus an input device. The results are usually displayed on the Command Window or in a Figure window. Thus, the computer screen is the output device. Another way to give input data is by using a file where we have somehow stored the input data. This file can be created by MATLAB or by any other computer program. This last option allows data exchange between MATLAB and any other software package. For example, a spreadsheet such as Excel might exchange data with MATLAB, so they can be processed and visualized. In the same way, data generated in MATLAB can be used by other programs.

An obvious advantage of saving data in a file is that we can use it in a later MATLAB session or we can use it in another computer or send it to another user.

6.5.1   File Opening and Closing

To read or write data to a file we have first to open it. MATLAB can open a file using the instruction fopen that has the format

fid = fopen (file name, permissions)

Here, file name is the file name and it must exist in order to open it. If we want to write to a non-existing file, the file is created and then the data is written in. Permissions is a variable that specifies how the data is written in. The permission codes available are given in Tables 6.3 and 6.4. We can open as many files as we wish. fid is the handle that is used to identify the file. The handle values start with 3. Handle values from -1 to 2 have a special use in MATLAB and they are given in Table 6.4.

After using the data from a file, we must close it. The instruction to close files is fclose. The format is:

status = fclose(fid)

Table 6.3: Permit codes to open files

Permission	Action
'r'	Opens the file to read. The file must already exist. If it does not exist, it sends an error message.
'r+'	Opens the file to read and write. The file must already exist. If it does not exist, it sends an error message.
'w'	It opens the file to write. If the file already exists, it deletes its contents. If it does not exist, it creates it.
'w+'	It opens the file to read and write. If the file already exists, it deletes its contents. If it does not exist, it creates it.
'a'	It opens the file to write. If the file already exists, it appends the new contents. If it does not exist, it creates it.
'a+'	It opens the file to read and write.

Table 6.4: File handles

Handle fid	Meaning
-1	Error when opening the file. Usually, when we want to read from a non-existent file.
0	Standard input. Usually, the keyboard is always open to write ('r').
1	Standard output. Usually, the Command Window. Always open with permission to add data at the end of the existing one.
2	Standard error. Always open with permission to add data at the end of the existing one.

The variable fid has the handle's value for the file we wish to close. status is another handle that indicates if the file was successfully closed. status = 0 indicates that it was closed, while status = -1 indicates that the file was not closed.

Example 6.17 Use of fopen.

Let us suppose that we want to open a file called Example_six_17.txt. Then we use:

```
handle_file = fopen('Example_six_17.txt', 'r');
handle_file
handle_file =
-1
```

The value of handle_file is -1 because the file does not exist. If we use 'w' instead of 'r', we get:

```
handle_file = fopen('Example_six_17.txt', 'w');
handle_file
handle_file =
3
```

The new handle value is 3 indicating that before writing to file Example_six_17.txt, it was created because it did not exist before. We now create another file data.txt with:

```
handle_file = fopen('data.txt', 'w')
handle_file =
4
```

The handle for this file is 4 because handle values are sequentially assigned. Using the instruction dir we can see all the files currently open:

```
dir
.  Example_six_17.txt
.. data.txt
```

We can use any word processor, such as Office Word, the Notepad or the WordPad, and even we can open a text file with the MATLAB editor, to see their contents.

To create a file in another directory different from the one we are working on we only have to specify the path. For example,

```
File_name = fopen('C:\new\file.txt', 'w')
```

File_name= 5

To close a file we use the instruction fclose whose format is

status = fclose(fid)

fid is the handle corresponding to the file we wish to close. The value of the variable **status** indicates if the file was closed successfully (**status** = 0). If this was not the case and the file was not closed, MATLAB sends an error message to let us know that the file could not be closed. Before closing MATLAB, it is convenient to close all files open during the session. This avoids losing the data stored in them during the session.

6.6 Writing Information to a File

The simplest way to write information to a file is with the instruction fprintf. We have used this instruction before to display output data in the computer screen through the MATLAB Command Window as:

fprintf ('Display this text to the screen')
Display this text to the screen

To write to a file we can use this same instruction. Let us create two new files with,

handle1= fopen('file1.txt','w');
handle2= fopen('file2.txt','w');

To write to a file we need to use its handle as in:

fprintf (handle1, 'We write here to file1.txt \n');
fprintf (handle2, 'We now write here to file2.txt \n');

To see what it is written in them we can open them with the Notepad but first we close them with:

fclose(handle1);
fclose(handle2);

6.6.1 Reading and Writing Formatted Data

MATLAB allows users to read and write formatted data to a file. We show the procedure with an example.

Example 6.18 Writing formatted data

Suppose we want to write data about the planets in the solar system. The data we wish to write is:

1) Name   10-character string.
2) Position   2-digit number.
3) No. of moons   2-digit number.
4) Diameter   10-digit number.

If any data is less than the characters indicated, the remaining characters are filled with blank spaces. The elements of each item are stored in an array.
    The data is:

```
Planet_name= ['Mercury';'Venus ';'Earth '; 'Mars ';...
'Jupiter';'Saturn ';'Uranus '; 'Neptune';'Pluto ']
Position= [ 1; 2; 3; 4; 5; 6; 7; 8; 9];
No_of_moons=[ 0; 0; 1; 2; 63; 34; 21; 13; 3];
Diameter_in_km=[ 4880; 12103.6; 12756.3;...
6794; 142984; 120536; 51118; 49532; 2274];
```

Note that we are writing the data as column vectors. To see the data for any planet we write:

```
fprintf ('%s \n %g \n%g\n %g\n', Name(2, :), Position(2, :), ...
No_of_moons(2, :), Diameter_in_km(2, :));
```

```
Venus
2
0
12103.6
```

To write this information to a file we use the following script

```
%File Example_6_18.m
%
handle_planets = fopen('Planets.txt', 'w');
for i = 1: length(Position);
fprintf(handle_planets,'%7s ,%5d ,%2d ,%2d \n', ...
Planet_name(i, :), Position(i, :), ...
No_of_moons(i, :), Diameter_in_km(i, :));
end
fclose(handle_planets);
```

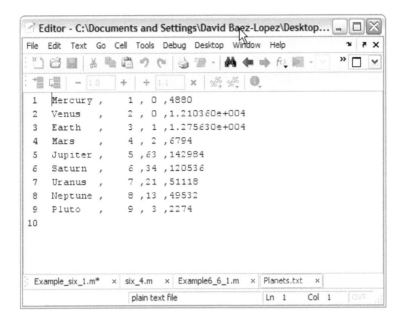

Figure 6.5: Contents of **Planets.txt.**

When we open the file Planets with the WordPad, we see the information shown in Figure 6.5.

We can now read the data in a file with a script or a function written in m-language. We have to check for several things when reading this data:

1. If we have gotten to an end_of_file which has the variable feof.
2. Read in each string with fscanf and assign it to its corresponding field.
3. Close the file after we find the feof.

The m-file is:

```
% This is file read_data.m
%
handle_data = fopen('Planets.txt', 'r')
% We define the names of the column vectors.
Names = [ ];
Positions = [ ];
Moons = [ ];
Diameters = [ ];
while  feof(handle_data)

% Read the planet name
```

```
stringg = fscanf(handle_data, '%7c', 1);
Names = [Names; stringg];
comma = fscanf(handle_data, '%1c', 1);

% Read the position
number = fscanf(handle_data, '%5d', 1);
Positions = [Positions; number];
comma = fscanf(handle_data, '%1c', 1);

% Read the number of moons
number = fscanf(handle_data, '%5d', 1);
Moons = [ Moons; number];
comma = fscanf(handle_data, '%1c', 1);

% Read the diameter
number = fscanf(handle_data, '%12e', 1);
Diameters = [ Diameters; number];

end_of_line = fscanf(handle_data, '%1c', 1);
end
fclose(handle_data);
```

The statement while checks for the end_of_file. fscanf search for the planet name characters using the format %7c. It reads the first seven characters including the blank spaces. If we have used instead the format %7s, only the characters are read and the blank spaces are ignored. The 1 after %7c means that only an element is read. In this way,

$$\text{stringg = fscanf(handle\_data, '\%7c', 1)}$$

does the following: reads an element of the open file which has the handle **handle_data** and place it in the variable stringg.

The line **comma = fscanf(handle_data, '%1c', 1)** indicates that after reading the first variable, a comma is read. We do not do anything with the comma but we need to read it. Otherwise it is read by the next instruction fscanf.

For the variables Positions, Moons, and Diameters, we need to read in a numerical value. For this we use

$$\text{number = fscanf(handle\_data, '\%5d', 1);}$$

After reading the data, we arrive at the end of the line. This is read with a special character

$$\text{end\_of\_line = fscanf (handle\_data, '\%1c', 1)}$$

Again, we do not need this character but we need to read it, for the same reason we did with the comma.

To see how this file works, we run the file and see the variables,

read_data
who

We now see the data

Names
Names =
Mercury
Venus
Earth
Mars
Jupiter
Saturn
Uranus
Neptune
Pluto

Positions

Positions =
1
2
3
4
5
6
7
8
9

Moons
Moons =
0
0
1
2
63
34
21
13
3

Diameters
Diameters =
1.0 e+005 *
0.0488
0.1210
0.1276
0.0679
1.4298
1.2054
0.5112
0.4953
0.0227

As we can see, these are the values entered before.

6.6.2   Reading and Writing Binary Files

So far, we have used alphanumeric data that can be read by any text processor besides the MATLAB editor. This type of data is called ASCII data. A disadvantage of this type of data is the size of the files when we handle a large amount of data. An alternative way is to store data in a binary format. This is a more efficient way to store information. Unfortunately, information stored in a binary format cannot be read by a text processor.

To write and read in a binary format we use the instructions fwrite and fread, respectively. The format for fwrite is:

$$\text{count} = \text{fwrite(handle , A, 'precision')}$$

The variable A contains the data to be written. count is the number of elements that were successfully written. handle is the handle for the opened file. precision gives information about how we want to store the information. Some of the options for this variable are given in Table 6.5.

Example 6.19 Reading and writing binary data

Let us suppose that we want to write the following data in binary format:

$$A = \begin{bmatrix} 57 & 10 \\ 14 & 75 \end{bmatrix} \qquad B = 27, \qquad C = \text{'MATLAB'}$$

First we open the file

    fid_binary = fopen('binary.dat', 'w');
    fid_binary = 3

To write we use

Table 6.5: Options for variable precision

Option	Meaning
'char'	8-bit characters. Used for text.
'short'	16-bit integers. Integer numbers in the range from -215 to 215-1.
'long'	32-bit integer (2's complement). Integers in the range from -231 to 231-1.
'ushort'	Unsigned integer (16 bits).
'uint'	Unsigned integer (32 bits).
'float'	Single precision floating point real numbers (32 bits).
'double'	Double precision floating point real numbers (64 bits).

fwrite(fid_binary, A, 'double')
ans =
4

The answer is 4 indicating that 4 elements were written. These elements belong to matrix A. We now write in B,

fwrite(fid_binary, B, 'short')
ans =
1

This time the result is a 1 because B has only one element. We now continue with C,

fwrite(fid_binary, C, 'char')
ans =
6

The answer is 6 because the word MATLAB has 6 elements. We now close the file

fclose (fid_binary)

Now we try to read the file with the WordPad and we get the results shown in Figure 6.6. As we can see, the data is not what we wrote in because it is in binary format and not in ASCII format, thus trying to read it with a text processor does not display the information stored in it.
To read the data, first we open the file

fid_bin = fopen('binary.dat', 'r')
fid_bin = 3

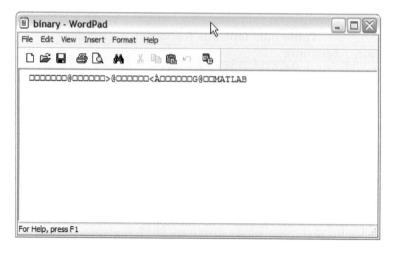

Figure 6.6: WordPad window showing the contents of **binary.dat**.

And now we use fread, whose format is:

[A, count] = fread(fid_binary, [2, 2], 'double')

Where A is the matrix name and count is the number of elements (4 in the case of matrix A). If A were an 8×4 matrix, then instead of [2, 2] we should write the size [8, 4]. The results are:

[A, count] = fread(fid_bin, [2, 2], 'double')

```
A =
   57   10
   14   75

count =
4
```

Data must be read in the same order and with the same format that it was written. To read out the remaining variables we use:

B = fread(fid_bin, [1], 'short')
```
B =
   27
```

C = fread(fid_bin, [6], 'char')
```
C =
77
65
```

```
84
76
65
66
```

Note that C is an ASCII string in a column vector. To change it to a row vector in characters we transpose it and then use setstr as in:

```
C = setstr(C')
C =
MATLAB
```

Finally, we close the file,

```
fclose(fid_bin)
ans = 0
```

This value indicates that the file was successfully closed.

6.7 Passing Data between MATLAB and Excel

A software package used in engineering, science, and finance is Excel. Excel and MATLAB can read and write data to files. In this section we show how such files can be used by any of the packages. For example, MATLAB can write data separated by commas in files with extension csv, for comma separated values, and then read by Excel, and vice versa.

6.7.1 Exporting Data to Excel

To show how we can export data from MATLAB to Excel we have the following example. Let us consider the following data about the five countries with the largest territories in square miles in the American continent together with their capital cities:

```
Canada, Ottawa, 3849660
United States of America, Washington D.C., 3787319
Brazil, Brasilia, 3300410
Argentina, Buenos Aires, 1073596
Mexico, Mexico D.F., 759589
```

This data is written by MATLAB to file countries.csv in the same way as we did in Section 6.6. The following file opens the file countries.csv, writes the data, and closes the file. We save this file to the directory Chapter 6.

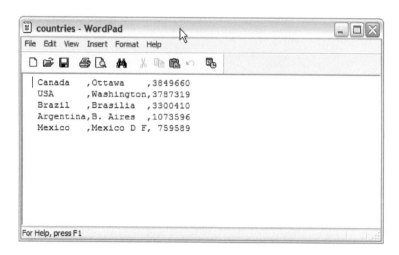

Figure 6.7: Data in the file **countries.csv**.

Country = ['Canada '; 'USA '; 'Brazil '; 'Argentina'; 'Mexico ']
Capital=['Ottawa '; 'Washington'; 'Brasilia ';'B. Aires ';'Mexico D F']
Size = [3849660; 3787319; 3300410; 1073596; 759589]
handle = fopen('countries.csv', 'w')
for i = 1: 5
    fprintf(handle, '%10s, %10s, %7d \n',...
    Country(i, :), Capital(i, :), Size(i, :))
end
fclose(handle)

Now we open the file countries.csv and we see the contents shown in Figure 6.7. Note that it is in comma separated values format. The commas are called separators or delimiters.

Now we proceed to open the file with Excel. Since the file was not created by Excel, it has to be imported to Excel. The Import Wizard is automatically open. It consists of three windows. The first window is shown in Figure 6.8(a). Here we indicate that the data is delimited by commas as indicated there. After pressing the Next button the second window for the Import Wizard opens and here we indicate that the data is delimited by commas as shown in Figure 6.8(b). Finally, when we press the Next button we get to the third window in the Import Wizard. Here we select the columns we want to import and set the data format, as shown in Figure 6.8(c). Finally, the data is shown in Excel as can be seen from Figure 6.9.

Another instruction used in MATLAB to write only numerical data to a file and read it by Excel is the instruction csvwrite ('file_name', m). For example, for the matrix A given by

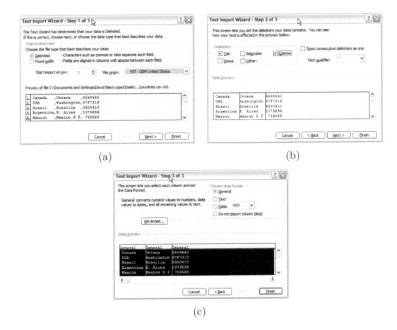

(a)   (b)

(c)

Figure 6.8: (a) **Part 1 of the Import Wizard. Here we indicate that the data is separated by commas or tabs. (b) Part 2 of the Import Wizard. Here we indicate that the data is separated by commas. (c) Part 3 of the Import Wizard. Here we select each column and set the data format.**

A = [1957 10 5; 1950 10 8;1989 5 10]
A =
   1957   10   5
   1950   10   8
   1989   5   10

We can write this matrix to a file list.csv with

csvwrite ('list.csv', A)

If we open this file with the Wordpad we obtain the window of Figure 6.10. There we see that the data is already separated by commas.

Since Excel can read data from files with extension csv, it readily opens the the file as shown in Figure 6.11.

6.7.2 Exporting Excel Files to MATLAB

The instruction csv also allows numerical data transfer from Excel to MATLAB. To show how this can be done, in Excel define the matrix A given by

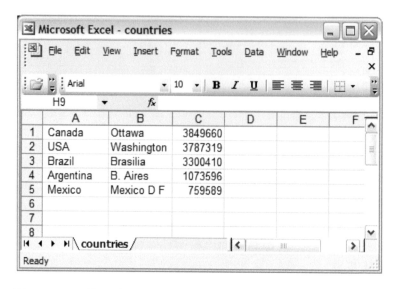

Figure 6.9: Excel window with data created in MATLAB.

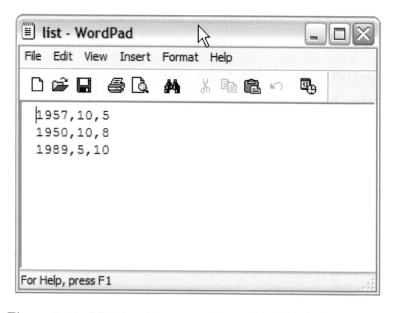

Figure 6.10: Matrix **A** in a **csv** file read with the Wordpad.

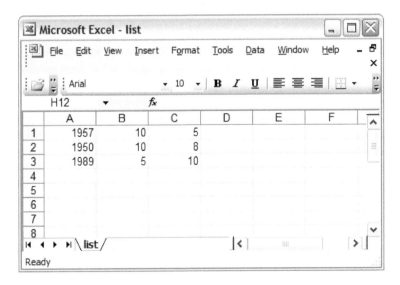

Figure 6.11: Matrix *A* in Excel in the file **list.csv**.

A =

1	10
100	1000
2	4

The matrix in Excel is shown in Figure 6.12. We store in a file **Numbers.csv**. Now, from the MATLAB Command Window we use the instruction csvread as

csvread ('Numbers.csv')

to obtain the data in MATLAB as shown in Figure 6.13. MATLAB can also read data from files with extension xls. To describe the procedure we use the Excel data shown in Figure 6.14. This data is saved in the file years.xls in the current directory for the MATLAB session. Now, in MATLAB we look at the Current Directory window and locate the file years.xls. We just double click on this file and then the Import Wizard opens requesting the variables to be imported. Selecting the variable to be imported (see Figure 6.14), it is displayed in the right-hand window and then we click on the **Finish** button to end the importing. In the Workspace window appears the variable which we can now use as any other variable created in MATLAB. This is shown in Figure 6.15.

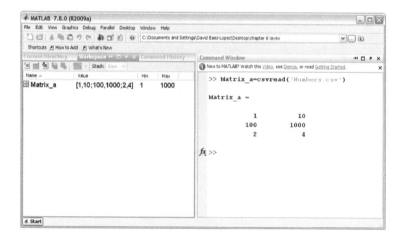

Figure 6.12: Matrix A imported to MATLAB.

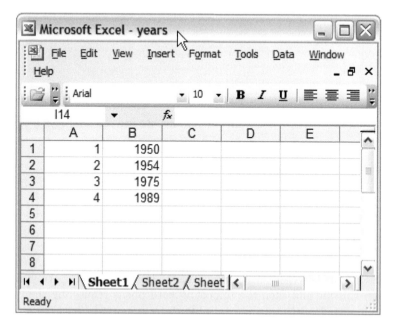

Figure 6.13: Data in an Excel file.

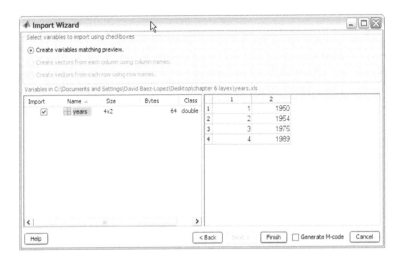

Figure 6.14: Import wizard for Excel files.

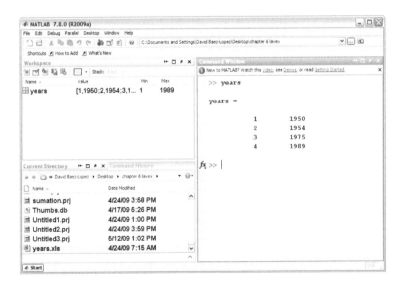

Figure 6.15: MATLAB window showing the file **years.xls** in the Current Directory, the variable **years** at the Workspace window, and the values of **years** at the Command Window.

6.8   Deployment of MATLAB m-files

MATLAB allows deployment of MATLAB files so they can be distributed and used by other users without having a MATLAB license. These deployed m-files can be used from EXCEL, .NET, Java, or as stand-alone executable files. MATLAB has a tool called Deployment tool that guides us in the making of executable files. There are four steps that have to be followed to obtain an executable file from an m-file. These are:

1. Create a project.
2. Add the files.
3. Build the executable file.
4. Pack the project.

Now we describe each one of the steps:

1.   We start the Deployment Tool from the MATLAB Start menu as shown in Figure 6.16. There we select

$$\text{MATLAB Start} \rightarrow \text{MATLAB Compiler} \rightarrow \text{Deployment Tool}$$

This opens the deployment interface shown in Figure 6.17. Then we select the icon for New Project and what type of executable we wish to make.
2. We now add the files we need in our project. In the file, Main function we add the main program and in the Other functions folder we add all the functions associated to the main program. If there are any C/C++ files we add them to the folder C/C++.
3. Once we have the required m-files, we proceed to build the project. Here we create the executable file.
4. Once we have the executable files, we pack them together with the MATLAB functions. This step creates a file with all the executable files for our m-files and functions and the MATLAB compiler named MATLAB MCR. This compiler contains all of the necessary functions required for a correct execution of the m-files and functions. MCR is an acronym for **MATLAB Compiler Runtime**.

We show with an example the procedure to create and pack an executable file.

Example 6.20 Deployment of the m-file from Example 6.7

We use the m-file from Example 6.7. We create an executable file and then we pack it. The file reads in a matrix and then sums up the elements of the matrix. The m-file is repeated here for convenience, but we have renamed it six_20:

```
% File six_20.m
% This m-file evaluates the sum of
% the elements of a matrix n x m.
n = input ('Enter the number of rows\n');
m = input ('Enter the number of columns\n');
% Reads in the elements
% Initialize the sum.
sum = 0;
% Reads in the matrix elements and adds them.
for i = 1: n
    % Reads in the elements of row i and adds them.
    for j = 1: m
        fprintf('Enter the matrix element %g,%g', i, j);
        a(i, j) = input(' \n');
        sum = sum+a(i, j);
    end
end
fprintf('The total sum is %g\n', sum)
```

We now start the deployment procedure. We number the steps as in the procedure outlined above and repeated here:

1. We start the deployment tool with

MATLAB Start→ MATLAB Compiler → Deployment Tool

as it is shown in Figure 6.16. This opens the Deployment Tool window of Figure 6.17. The toolbar of the Deployment Tool is shown in Figure 6.18 and the two important icons are the Build the project icon and the Package the Project icon shown in Figure 6.19. In this window we open a new deployment project for a Windows stand-alone application as shown in Figure 6.20 where we have given the name to our project Sum_of_elements.prj.

2. We now drag the file six_20.m to the Main function folder as shown in Figure 6.21.

3. We build the executable by clicking on the Build the Project icon. In the Deployment Tool Output window we see the message

mcc -F 'C:\chapter_ 6\sum_of_elements.prj'

After a few seconds we get a message that the executable file was successfully made. If this is not the case, we have to correct the problems and again build the executable.

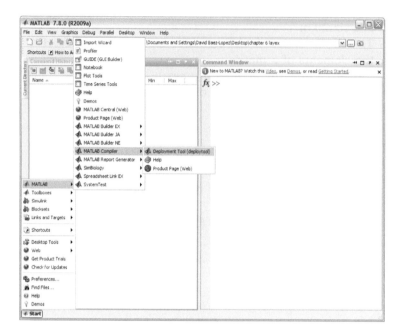

Figure 6.16: Starting the **Deployment** tool.

4. We pack the executable file together with the MCR for deployment. We do this by clicking on the Package the Project icon. After this process ends, we have a file Sum_of_elements.prj and a folder Sum_of_elements. **These two products of the deployment procedure are the parts that are distributed to the end user.** The Sum_of_elements folder contains the folders distrib and src.The folder distrib contains the files:

   a. _install.exe

   b. Sum_of_elements.exe

   c. Sum_of_elements_pkg.exe

5. We are now ready to distribute our package

The file _install.exe includes the file MCR which is installed when we unpack the project in the destination computer. Then we execute the file

Sum_of_elements_pkg.exe

which installs all the remaining parts to execute the file six_20 in an equipment without a MATLAB license.

Finally, we execute the file Sum_of_elements.exe to open the window requesting the input data. A run is shown in Figure 6.22.

As we see, it is very easy to deploy MATLAB programs to use in a computer that does not have a MATLAB license.

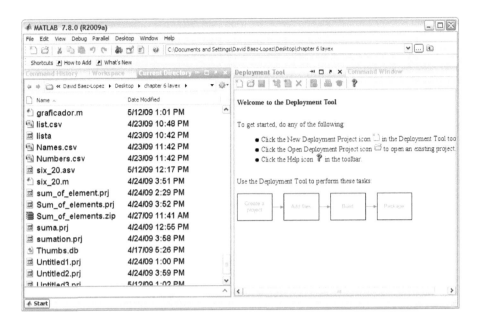

Figure 6.17: Deployment tool.

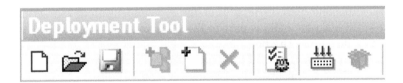

Figure 6.18: Deployment tool toolbar.

Figure 6.19: Icons for building the project and package it.

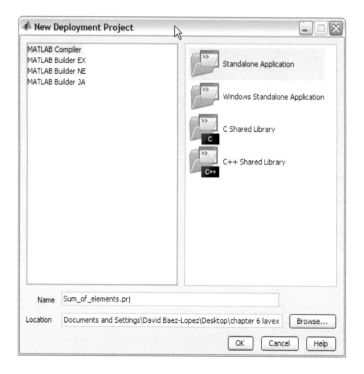

Figure 6.20: Selection of compiler and project name.

6.9 Publishing m-files from MATLAB

A very important part in programming is documentation. Sometimes this may be either a very tedious part in the programming process or a very easy one depending upon the programmer's style. Fortunately, in the case of m-language programming, MATLAB has a tool to create documentation once a program has been finished. This is known as publishing and the end result is a document that can be created in Word, HTML, XML, Latex, or Power Point. Furthermore, it is possible to run the program documented from the published file. We show the procedure with an example.

In order to learn the way an m-file has to be structured to publish it first we have to describe cell programming.

6.9.1 Cell Programming

The MATLAB editor has the option to create cells. These cells are useful to run portions of the m-file and to create sections in the publishing process. A cell is a portion of an m-file having certain characteristics.

A cell is composed of the following parts:

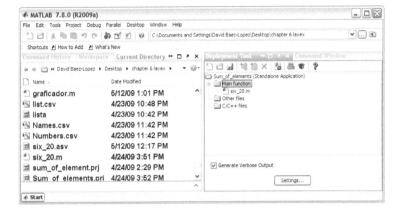

Figure 6.21: File **six_20** in the **Make function** folder.

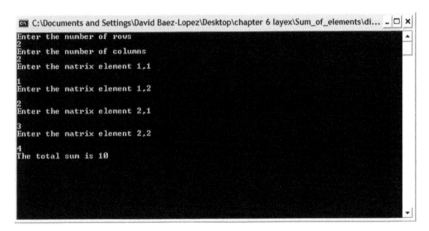

Figure 6.22: A run for the file packaged.

1. A beginning row which starts with a double percent sign followed by a space and a title text.

2. Comment lines that start with a percent sign followed by a space and text which is the body of the documentation.

3. Equations written in latex style, and finally:

4. MATLAB instructions.

We show the procedure with an example.

Example 6.21 Plotting of a sine function with cells

Let us suppose that we want to plot the sine function from 0 to 2π and then we want to modify the plot. To plot the function we use

```
x = 0: 0.01: 2*pi;
y = sin(x);
plot(x, y)
```

Once the function is plotted we add title, legends to the axes and a label with:

```
xlabel('x-axis')
ylabel('sine wave')
title('Plot of sin x')
```

Now we create the m-file using cells. In the first cell we place the first part of the m-file and in the second cell the m-file where we add the text part. We add comments to the m-file to make it self-explanatory. In the comment lines we leave a blank space between the percent sign % and the beginning of the text. The following m-file is in cell form:

```
%% Example of m-file using cells
%% Plot of a sine wave
% Here we plot a sine wave going from 0 to
2\pi
% As we know we have to create a vector for x values
% and then a vector for y values. The vector y has the information
% for the sine wave values
x = 0: 0.01: 2*pi;
y = sin(x);
plot(x, y)
%% Adding text information to a plot
% We add a title with title.
% We add a text to the x axis with xlabel.
% Finally, we add a text to y axis with yaxis.
%
xlabel('x-axis')
ylabel('sine wave')')
title('Plot of sin x')
```

The editor window is shown in Figure 6.23. We see that each cell is separated by a line and that each cell has different background color. We also note that there is a line below the toolbars that says:

This file uses Cell Mode. For information

This indicates that we can run the file in cell mode. To run the file in this mode, from the main menu we select Cell→Evaluate Current Cell and Advance

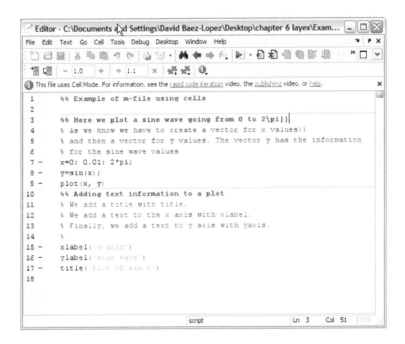

Figure 6.23: m-file divided in cells.

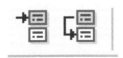

Figure 6.24: Icons to run an m-file in cell mode. Single cell run (left) and run the cell and advance to following cell (right).

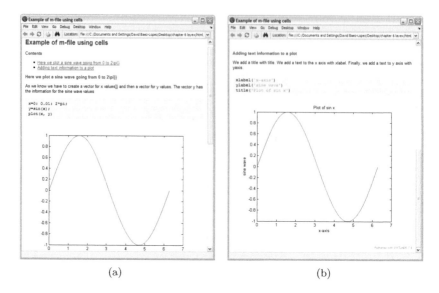

(a)   (b)

Figure 6.25: HTML document. (a) Top part, (b) Bottom part.

and this runs the m-file a cell at a time. To run in cell mode we can also use
the cell icons the second toolbar in the editor, shown in Figure 6.24.

After running each cell we see the result of the second cell is the sine
function plot, and the last cell result is the same plot with a title and with axes
labels. The first cell do not display anything because there are no instructions
in that cell. We now proceed to publish this m-file.

6.9.2   Publishing m-files

Now that we have the m-file in cell mode we can proceed to publish it. The
result is a document in the format selected. The first step is to run the
publisher tool. This is done from the MATLAB m-file editor. We can choose
publishing in Word, HTML, XML, Latex, or Power Point. The default option
is the HTML format. If we wish to change to any of the other formats we can
do so in the MATLAB preferences from the File menu. For the example we
choose the default HTML format. From the main menu select File→Publish
to HTML. This will start the publishing process. After a few seconds we get
the HTML document shown in Figure 6.25.

We see in the HTML document that it has

1. A title,
2. A table of contents,
3. MATLAB instructions,
4. Text explaining the instructions, and

5. Plots.

We now describe each part of the document.

1. The title "Example of m-file using cells" is the first line of the first cell (the first cell has only the title of the document).

%% Example of m-file using cells

2. The table of contents is formed by the first line of each cell. That is, the lines with a double percent sign,

%% Adding text information to a plot
%% Plot of a sine wave

3. The MATLAB instructions are, for the second cell,

```
x = 0: 0.01: 2*pi;
y = sin(x);
plot(x, y)
```

and for the third cell

```
xlabel('x-axis')
ylabel('sine wave')
title('Plot of sin x')
```

4. The text for each cell is the comment lines. For the second cell:

% As we know we have to create a vector for x values
% and then a vector for y values.
% The vector y has the information
% for the sine wave values.

For the third cell:

% We add a title with title.
% We add a text to the x axis with xlabel.
% Finally, we add a text to y axis with yaxis.

5. Finally, the plots are also displayed in the HTML document.

In the published document we can also include equations. They have to be written in a LaTeX format. Table 6.6 lists some of the more used MATLAB LaTeX formats to write equations. For example, to write

Table 6.6: Commands for MATLAB LaTex

Traditional	MATLAB LaTex
Equation	$$ equation $$
$\frac{a}{b}$	\frac{a}{b}
a^2	a^2
$a_{k,n}$	a_{k,n}
α^2	\alpha^2
$\sqrt{(a+b)}$	\sqrt{(a+b)}
$\int(a+b)dt$	\int{(a+b)dt}
a$\leq$ b	a\leq b
a$\geq$ b	a\geq b
\	\textbackslash

$$\int \sqrt{\alpha sin(t)}dt$$

We write

$$\% \ \$\$ \ \backslash int \backslash sqrt\{\backslash alpha\{sin \ (t)\}\}\backslash, \ dt \ \$\$$$

Some rules have to be followed. These rules are:

1. If a row has an equation, there must be an empty comment row above and below.

2. There must be at least a blank space between the percent sign and the dollar sign.

3. After a row with executable MATLAB instructions, the row has to start with a double percent sign, that is, a new cell has to start.

We now show an example to solve a quadratic equation.

Example 6.22 Publishing an m-file with equations

To solve the quadratic equation

$$ax^2 + bx + c = 0$$

We have the solutions:

$$x_{1,2} = \frac{-b \pm \sqrt{b^2 - 4ac}}{2a}$$

To publish these equations we have to write the m-file as

```
%% Solution of a second order equation
%% Introduction
% A second order equation of the form
%
% $$ ax^2 +bx+c = 0 $$
%
% has the solutions
%
% $$ x_{1} = \frac{-b + \sqrt{b^2-4ac}}{2a} $$
%
% $$ x_{2} = \frac{-b - \sqrt{b^2-4ac}}{2a} $$
%
%% Example
% As an example we solve the equation
%
% $$ 3x^2+6x-9 = 0 $$
%
% The data are then a = 3, b = -2, c = 7.
a = 3; b = 6;c = -9;
%% Solutions
% The solutions are
%
x1 = (-b+ sqrt(b^2- 4*a*c))/(2*a);
x2 = (-b- sqrt(b^2- 4*a*c))/(2*a);
%% Results
% Finally, we display the values of the roots.   %
fprintf ('x1 is a root of the second order equation %g\n', x1)
fprintf ('x2 is a root of the second order equation %g\n', x2)
```

We publish to HTML and we get the document shown in Figures 6.26 and 6.27.

6.10   Concluding Remarks

MATLAB has integrated a powerful programming language called m-language. This language allows users to produce complex programs in a very short time when we compare it with other programming languages as C, C++, Visual Basic, FORTRAN, among others. In the chapter we treated in detail several of the instructions needed to write a program in the m-language. The process was carried out through examples going from very simple ones to more complex programs. A treatment in input/output instructions was given, in particular, the case of reading to/from a file. Also the case of transferring information between MATLAB and Excel was treated. More advanced topics such as deployment of m-files to users not having a MATLAB license was also

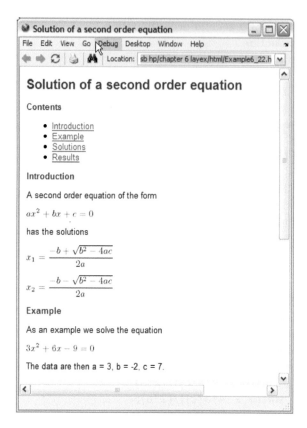

Figure 6.26: Top part of the published file in HTML format.

discussed and examples provide a good understanding of the topic. Finally, since program documentation, also known as publishing, is an important part of programming, MATLAB also provides a tool for documentation m-files. The process can publish to several formats, but we only used the one for an HTML document.

6.11   Exercises

Section 6.2

6.1 Write an m-file to find the solutions for the equation

$$e^{-x}$$

using an iterative method. Compare your result with the one obtained using the instruction **solve**.

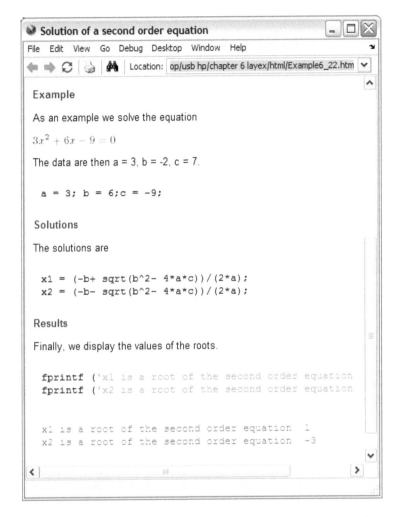

Figure 6.27: Bottom part of the published file in HTML format.

6.2 The characteristic equations for a MOS transistor are

$$I_D = K(V_{GS} - V_T)^2 \qquad \text{for} \qquad V_{DS} \geq V_{GS} - V_T$$

$$I_D = K\{2(V_{GS} - V_T)V_{DS} - VDS^2] \qquad \text{for} \qquad V_{DS} < V_{GS} - V_T$$

Make a plot of I_D vs. V_{DS} using V_{GS} as a parameter.
Use $K = 0.0025$, $V_{GS} = 0.01, 0.02,$ and 0.03.

6.3 Write an m-file to accept an input number. If the number is less than 10, the m-file must write "The data is correct". If this is not the case, then write "The data is not correct". Use the instruction while.
The program must check if the data is a numerical or a string.

6.4 A candy vending machine sells three types of candy. Let the candies be named A, B, and C, and the prices are $1.00, $2.00, and $3.00, respectively. Write an m-file that counts the coins inserted in the machine and computes the amount. If the amount is correct it dispenses the product selected.

6.5 To the m-file of Exercise 6.4 add the instructions required to give change if the coin inserted is a $5.00 or a $10.00 coin.

6.6 Write an m-file that finds the first three integer numbers whose square is greater than 2000.

Section 6.3

6.7 Write a function that computes the area and volume of a cylinder. The input data is the radius and the height.

6.8 Write a function that finds the area of a parallelogram with a base of 12 centimeters and a height of 5 centimeters.

6.9 Write a function that evaluates the mean of three numbers read in.

Section 6.4

6.10 Write two functions, the first of them to plot the path of a ball thrown with an angle θ. The second function should find the optimum angle θ for a maximum distance. Use global variables.

6.11 Write a script that finds the value for the function $\sin\theta$. Start with the Taylor series expansion

$$\sin x = \sum_{n=0}^{\infty} \frac{x^n}{n!}$$

This expansion is only valid for small values of θ. Then, compare it with the result using the MATLAB function sin(x). Use global variables.

6.12 Write a recursive function to evaluate x^n. Use the fact that

$$x^n = x \cdot x^{n-1}$$

Section 6.5

6.13 Plot the functions $y = \cos(x)$, $z = y^2$. Plot the first function in Figure 3 and the second at Figure 54.

6.14 Plot the functions $y = \sin(x)$, $z = \cos(x)$, $w = \exp(x)$ in the rank from -1 to 1 using handle_plot=plot(x, y, x, z, x, w) to obtain the handles of each of the three traces. Then, add a text with handle_text=ylabel('y-axis'). Repeat for the x-axis and for the title of the plot. Obtain the handle for the axes.

Section 6.6

6.15 Using information of the football league, write in a file the information of the ten best players. Then, read in this information to a MATLAB file. The information required is: Name, team, player's number, and touchdowns.

6.16 Write file from Exercise 6.15 in a binary format. Open the file with the text editor and change any of the fields. Open the file with MATLAB and see how that data was changed.

Section 6.7

6.17 Write the following data in an Excel spreadsheet and read it from MATLAB and plot it.

Time	Cost
0	-2
1	-2.5
2	2.4
2.5	2.3
3.3	2.2
8	1
12	0
17	4
23	7
27	7.5

6.18 Write from MATLAB a file with the largest-population countries in Europe and then export it to Excel.

6.19 Create a csv file with the numbers from 1 to 10 and then read it from Excel and MATLAB.

Section 6.8

6.20 Obtain the executable file for a file that finds the solution for a second-order equation.

6.21 Make an executable file that solves a set of two linear simultaneous equation.

6.22 Make an executable file that plots the solution of the differential equation

$$mL\ddot{\theta} + BL\dot{\theta} + w\sin\theta = 0$$

This equation describes the movement of a pendulum. Here L is the pendulum length, B is the damping coefficient, w is the weight, and θ is the angular displacement.

6.23 Write a file that plays a sound of 8 KHz. Make the executable that plays this sound. The file sound should be in a wav file.

Section 6.9

6.24 Publish the results of Exercise 6.22.

6.25 Publish the solution of a parabolic throw.

6.12   References

[1] H. Moore, MATLAB for Engineers, 2nd Edition, Prentice Hall, Inc., Piscataway, NJ, 2008.

[2] M. E. Herniter, Programming in MATLAB, Brooks/Cole Thomson, Pacific Grove, CA, 2001.

[3] S. Nakamura, Numerical Analysis and Graphic Visualization with MATLAB, Prentice Hall, Inc., Piscataway, NJ, 1995.

[4] MATLAB Programming Fundamentals, The MathWorks, Inc., Natick, MA, 2009.

[5] MATLAB Compiler User's Guide, The MathWorks, Inc., Natick, MA, 2009.

[6] MATLAB Report Generator User's Guide, The MathWorks, Inc., Natick, MA, 2009.

Chapter 7

Graphical User Interfaces

A graphical user interface (GUI) is the link between a sofware package and the user. In general, it is composed of a set of commands or menus, instruments such as buttons, by means of which the user establishes a communication with the program. The GUI eases the tasks of inputting data and displaying output data.

7.1 Creation of a GUI with the Tool GUIDE

MATLAB has a tool to develop GUIs in an easy and expeditious way. This tool is called Graphical User Interface Development Environment and better known by its acronym GUIDE. This tool can create a GUI empty window, add buttons and menus to add to our GUI, and windows to enter data and plot functions, as well as the access to the objects' callbacks. When we create a GUI with GUIDE, two files are created: a fig file which is the graphical interface and an m-file which contains the functions, the description for the GUI parts, and the callback.

A callback is defined as the action that realizes an object of the GUI when the user clicks on it or uses it. For example, when the user clicks on a button in a GUI, a program containing the instructions and tasks to be realized is executed. This program is called the callback. A callback is coded in the m-language.

7.1.1 Starting GUIDE

GUIDE can be started from the MATLAB window by clicking on the GUIDE icon in the MATLAB toolbar. It can also be started with File→ New→ GUI. Any of these choices opens the GUIDE Quick Start window shown in Figure 7.1. Here we can start a new GUI with an empty blank GUI where we can add and arrange the object for the GUI. We can also start a new GUI with

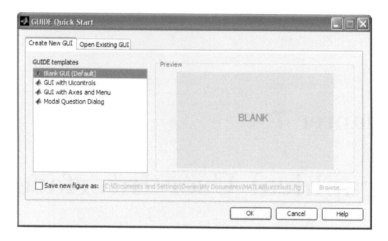

Figure 7.1: GUIDE Quick Start window.

Table 7.1: Important icons in the GUIDE toolbar

Property inspector	It refers to the properties of each object in the GUI. They include color, name, tag, value, and the callback among others.
Align Objects	It aligns the objects in the work window.
Toolbar editor	It creates a toolbar in the GUI.
M-file editor	It opens the MATLAB editor to edit the callbacks.

uicontrols, with a set of axes and menu already in the GUI, and finally, a GUI with question dialog buttons. Alternatively, we can continue a GUI previously started in the tab Open Existing GUI. If we select Blank GUI we get the work window of Figure 7.2. In this work window we see at the left a set of buttons or objects that can be used in the GUI. Each button has a function which is described by the button's name and it is self-describing. Additionally, the work window has a toolbar which has icons to open the Property Inspector, to align the objects in the GUI layout, to create a toolbar in the GUI, and to open the m-file editor where we can edit the callback for the elements in the GUI.

7.1.2   Properties of Objects in a GUI

Each object in the GUI has properties that can be edited with the Property Inspector. For example, for a Push button Figure 7.3 shows the Property Inspector with all the properties of this button. Some of the most common properties are shown in Table 7.2.

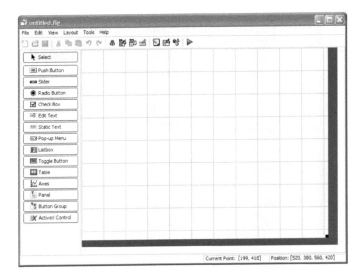

Figure 7.2: A blank GUI.

Table 7.2: Most used properties for objects in a GUI

Background color	Changes the background color of the object.
Callback	It contains the instructions to be executed by the object.
Enable	Activates the object.
String	Mostly used in the cases of buttons, Edit text boxes, and Static text boxes. It contains the text displayed in the object.
Tag	It identifies the object.

7.1.3 A Simple GUI

We show the procedure with a simple GUI. Let us suppose that we wish to create a GUI that plots a user-defined function. Thus, we need a text box to enter the function, two text boxes to enter the initial and final points in the plot, a set of axes to plot the function, and a button to run the GUI. Additionally, we can add some static texts to make the GUI and a button to close the GUI after we finish.

1. The first step is to open a Blank GUI work window.

2. We add the required objects. We start with two push buttons, three Edit text boxes, a set of Axes, and five Static Texts. The GUI is shown in Figure 7.4.

3. We can stretch each object to the desired size as we can see in Figure 7.5.

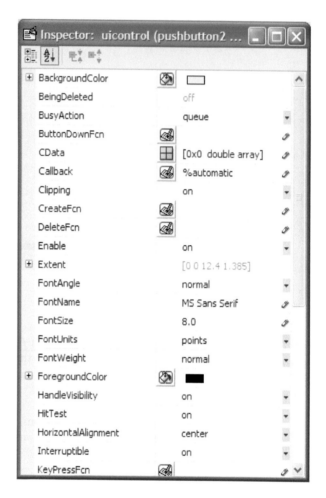

Figure 7.3: Property inspector.

4. We double-click on each one of the objects to open the Property inspector and change the String in the Static texts as shown in Figure 7.6. The Font weight property is changed to bold. For each of the remaining elements we clear the Strings.

5. For the Edit text box we change the Tag property to The_function. This is the variable name for the function to be plotted. For the remaining Edit text boxes we change the tags to Initial_x and Final_x. At this point the GUI looks as shown in Figure 7.7.

6. We change the Tag properties of the push buttons to Plot_function and CloseGUI.

7. We now need to edit the callbacks.

8. First we edit the callback of the Close button. We only need to add

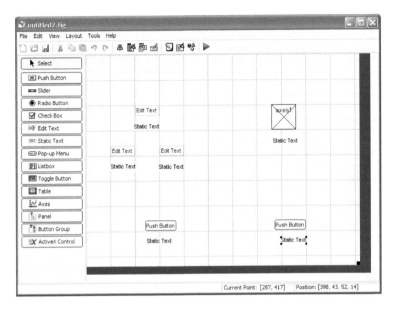

Figure 7.4: GUI with required objects.

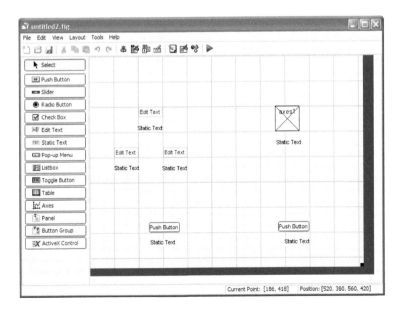

Figure 7.5: GUI with objects stretched to final size.

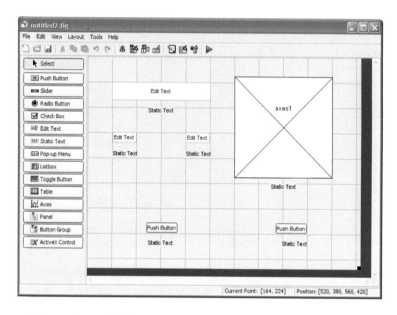

Figure 7.6: GUI with objects stretched to final size.

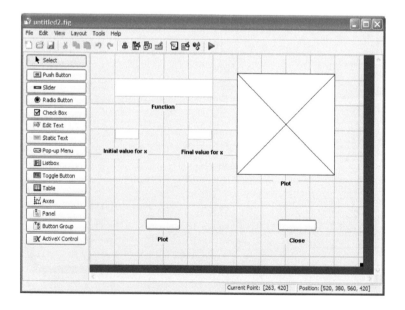

Figure 7.7: Final GUI layout.

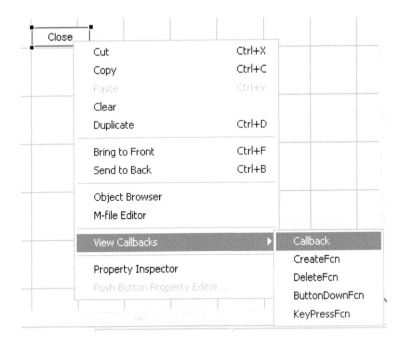

Figure 7.8: Menu to open and edit callbacks.

the instruction

$$close(gcbf)$$

which indicates to close the figure where the object is embedded. In this case the figure, that is the GUI, where the **push** button is located. To edit the callback we select this **push** button and click on the mouse's right-hand button to open the menu shown in Figure 7.8 where we select View callbacks→ Callback. This opens the m-file editor in the portion corresponding to this push button. The m-file is then as follows:

```
% — Executes on button press in CloseGUI.
function CloseGUI_Callback(hObject, eventdata, handles)
% hObject handle to CloseGUI (see GCBO)
% eventdata reserved - to be defined in a future version of MATLAB
% handles structure with handles and user data (see GUIDATA)
close(gcbf)
```

9. We now edit the callback for the button Plot. In this callback we have to read in the function we wish to plot as well as the lower and upper limits for the variable x, and finally, to plot it in the set of axes in the GUI. First, we open the callback for the button Plot by right-clicking on the push button

and selecting View callbacks→ Callback. This opens the m-file editor. To read
the information in the strings and make it amenable for calculations we use,
for example for the lower limit of x

```
lower_x_value = eval(get(handles.Initial_x, 'string'));
```

The instruction get fetches the string variable which is located in the object
with the handle Initial_x and the instruction eval converts it to a real variable.
A similar instruction applies for the upper x limit and for the function to be
plotted. That is,

```
lower_x_value = eval( get( handles.Initial_x, 'string'));
upper_x_value = str2num( get( handles.Final_x, 'string'));
y = eval( get( handles.The_function, 'string'));
```

10. Now, we are ready to get the plot. We now get the x axis points with

```
xx = [lower_x_value:0.2:upper_x_value];
```

11. The function to be plotted is now in the symbolic variable y. To be
able to accept x as a variable symbolic, we add the instruction syms x.
12. To evaluate the function at the set of points xx we substitute the
variable x with the point vector xx with

```
yb = subs(y,x,xx);
```

13. Finally, we plot the vector yb with

```
plot(xx,yb)
```

The final callback for the push button Plot is

```
% — Executes on button press in Plot_function.
function Plot_function_Callback(hObject, eventdata, handles)
% hObject handle to Plot_function (see GCBO)
% eventdata reserved - to be defined in a future version of MATLAB
% handles structure with handles and user data (see GUIDATA)
syms x
lower_x_value = eval( get( handles.Initial_x, 'string'));
upper_x_value = str2num( get( handles.Final_x, 'string'));
xx = [lower_x_value: 0.2: upper_x_value];
y = eval( get( handles.The_function, 'string'));
yb = subs(y, x, xx);
plot(xx, yb)
```

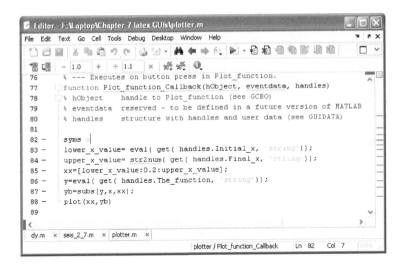

Figure 7.9: Callback for the **Plot** button.

The callback for the Plot button is shown in Figure 7.9.

To finish the GUI we add a toolbar. From the Tools menu in the GUI select Toolbar editor and in the window that it is opened we select the icons we wish to have in the toolbar. Once we are finished we close the GUI work window and to run the GUI we type Plotter in the MATLAB command window. Figure 7.10 shows a plot for the function exp(-x/20)*sin(x).

7.2   Examples

This section presents some examples introducing some of the other objects for the GUIs. The first example shows the sliding control and the pulldown menus for a GUI that converts temperature given in Fahrenheit degrees to Celsius and vice versa. The second example shows a GUI that calculates the derivative, the integral and the Fourier transform for a user-provided function. The last example is a GUI for the calculation of put and call options using the Black-Scholes function from the Financial Derivatives toolbox.

Example 7.1 Calculation of put and call options in finance

In Chapter 11 we show how to calculate call and put options for European options. There we show that we have to solve the Black-Scholes differential equation

$$\frac{\partial f}{\partial t} + rS\frac{\partial f}{\partial S} + \frac{1}{2}\sigma^2 S^2 \frac{\partial^2 f}{\partial S^2} = rf$$

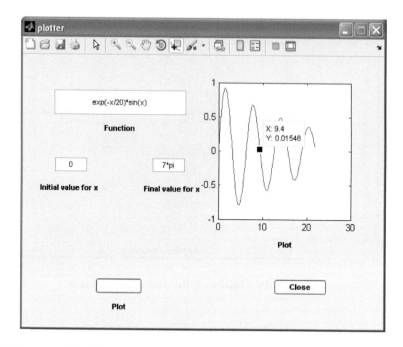

Figure 7.10: Plot of function exp(-x/20)*sin(x) from 0 to 7π.

whose solution is given by:

1. For the Call Option

$$c = S_0 N(d_1) K e^{-rT} N(d_2)$$

2. For the Put Option

$$p = K e^{-rT} N(-d_2) - S_0 N(-d_1)$$

where

$$d_1 = \frac{\ln(\frac{S_0}{K}) + (\frac{r+\sigma^2}{2})T}{\sigma\sqrt{T}}$$

$$d_2 = \frac{\ln(\frac{S_0}{K}) + (\frac{r-\sigma^2}{2})T}{\sigma\sqrt{T}} = d_1 - \sigma\sqrt{T}$$

Here $N(x)$ is the cumulative probability distribution function for a variable that is normally distributed with a mean of zero and a standard deviation of 1, $S0$ is the stock price at time zero, and K is the strike price. The function $N(x)$ is integrated into MATLAB as normcdf(x). In this example, we construct a GUI that has as input the stock price $S0$, the strike price K, the maturity time

T, the interest rate variation r, and the volatility σ. We use the Black-Scholes function from the Financial Derivatives toolbox that has the format

$$\textbf{[Call, Put]} = \textbf{blsprice(Price, Strike, Rate, Time, Volatility)} \qquad (7.1)$$

For example, for the data : stock price $S0 = 42$, strike price $K = 40$, interest rate $r = 10\%$, maturity time $T = 6$ months $= 0.5$ years, and a volatility $\sigma= 20\%$, we have

$$\textbf{[Call, Put]} = \textbf{blsprice (42, 40, 0.1, 0.5, 0.2)}$$

which gives the results for the call and put options as

Call =
4.7594
Put =
0.8086

The layout for the GUI to carry out this computation is shown in Figure 7.11. To this layout we change the strings for each Static text, each Edit text, and the Push buttons as shown in Figure 7.12. Then, we change the tags for the Edit text boxes with the first word in the name of the Static box next to each of them. That is, the Edit text next to Stock price has the tag equal to Stock, and so on. For the Static text boxes we also give the tag name in the same way. Then the top Static text box next to Call option we make the tag equal to Call and the other one has the tag equal to Put. For the push buttons we give the tags Calculate and Close. Now we edit the callback for the button Close as in the previous example. The callback for this push button is

```
% — Executes on button press in Close.
function Close_Callback(hObject, eventdata, handles)
% hObject handle to Close (see GCBO)
% eventdata reserved - to be defined in a future version of MATLAB
% handles structure with handles and user data (see GUIDATA)
close(gcbf)
```

The next step is to execute the Black-Scholes instruction from Eq. 7.1 when we click on the button Calculate. First we read the data from the Edit text boxes and then execute the instruction blsprice. Finally, we write the results to the empty Static boxes. The callback for the push button Calculate is now described:

1. First we read in the variables for the Black-Scholes instruction blsprice. We do this with the instructions eval and get. As we explained above, get reads the string from the Edit text and eval converts it the string to a numerical value assigned to the variable stock. To read the variable from the Edit box Stock we use then,

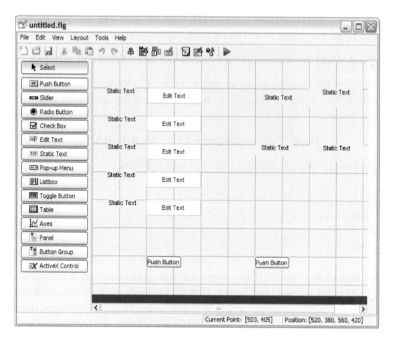

Figure 7.11: GUI layout.

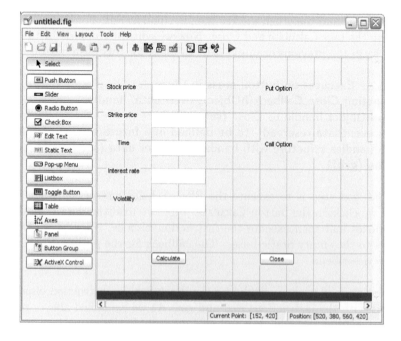

Figure 7.12: Final GUI layout.

stock = eval(get(handles.Stock, 'string'));

We read the five variables with

```
stock = eval( get(handles.Stock, 'string'));
strike = eval( get(handles.Strike, 'string'));
interest = eval( get(handles.Interest, 'string'));
maturity = eval( get(handles.Maturity, 'string'));
volatility = eval( get(handles.Volatility, 'string'));
```

2. Now, we make the calculation with the blsprice instruction with

```
[Call, Put] = blsprice(stock, strike, interest, maturity, volatility)
```

3. We write the results to the empty Static text boxes with

```
set(handles.call, v'string', Call)
set(handles.put,v'string', Put)
```

The complete callback for the push button Calculate is

```
% — Executes on button press in Calculate.
%
function Calculate_Callback(hObject, eventdata, handles)
% hObject handle to Calculate (see GCBO)
%
% eventdata reserved - to be defined in a future version of MATLAB
%
% handles structure with handles and user data (see GUIDATA
% )
stock = eval( get(handles.Stock, 'string'));
strike = eval( get(handles.Strike, 'string'));
interest = eval( get(handles.Interest, 'string'));
maturity = eval( get(handles.Maturity, 'string'));
volatility = eval( get(handles.Volatility, 'string'));
[Call, Put] = blsprice(stock, strike, interest, maturity, volatility)
set(handles.call, 'string', Call)
set(handles.put, 'string', Put)
```

Now we execute the GUI by clicking on the Run icon. The GUI looks as shown in Figure 7.13. Then we enter the values required and the results are shown in Figure 7.14.

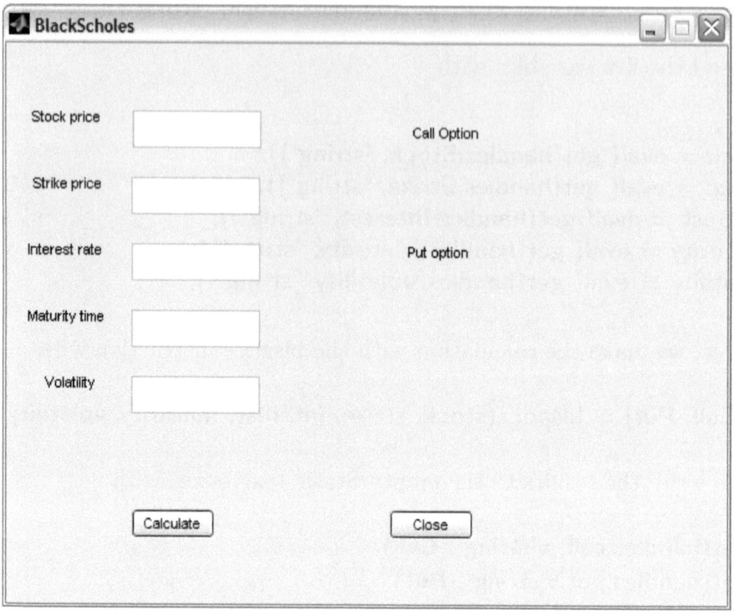

Figure 7.13: GUI for Black-Scholes calculation.

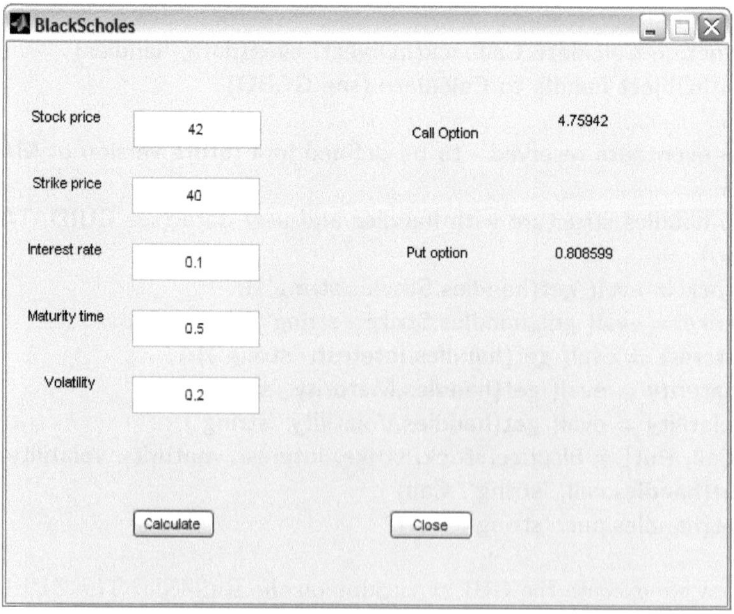

Figure 7.14: A run for the Black-Scholes GUI.

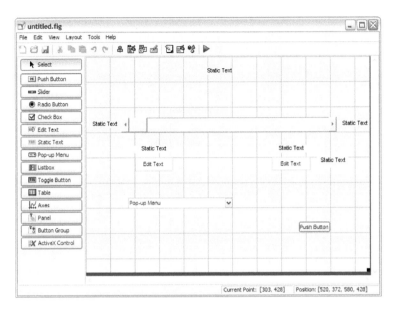

Figure 7.15: Initial layout for the GUI.

Example 7.2 Temperature conversion

Temperature conversion among the different scales, Celsius, Fahrenheit, and Kelvin, is possible if we know the conversion equations. They are available in any physics textbook and they are:

$$K = C + 273.15$$
$$K = 1.8(F - 32) + 273.15$$
$$C = 0.555(F - 32)$$
$$C = K - 273.15$$
$$F = 1.8(K - 273.15) + 32$$
$$F = 1.8C + 32$$

Now we create a GUI that realizes these conversion equations. We use the regular instructions to close the GUI and to read in data. To choose the conversion we use a Pop-up menu and for the input temperature data we use both an Edit text box and a slider. The initial GUI layout is shown in Figure 7.15. We change the strings for each of the GUI components so that they look as shown in Figure 7.16.

The Static text to the right of the result Edit box has a blank string. The string for the pop-up menu is set by clicking on the string property at the Property Inspector for the menu. This opens the String window for the pop-up menu where we add the information shown in Figure 7.17. We now change the tags for each of the components, as follows:

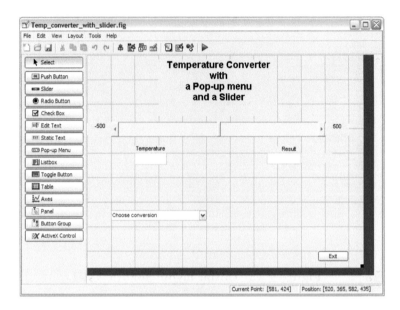

Figure 7.16: GUI layout with strings changed.

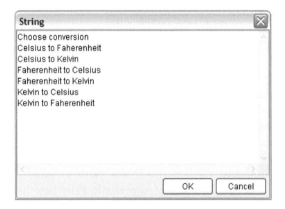

Figure 7.17: String for the pop-up menu.

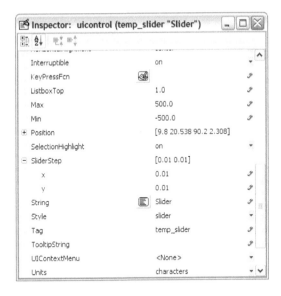

Figure 7.18: Property inspector for the slider.

GUI Component	Tag
Pop-up menu	temp_conv
Edit box below the text Temperature	input_temp
Edit box below the text Result	result
Push button	closeGUI
Slider	temp_slider

Before we begin to write the callback, we set the maximum and minimum values for the slider. They are set in the Property Inspector for the Slider which opens with a double click on the slider. For the Slider Step property we choose steps x = 0.01 and y = 0.01. These changes are shown in Figure 7.18.

We design the GUI in such a way that when the user chooses the conversion with the pop-up menu, the conversion takes place and the results are written to the result box. The callback we need to edit is the one corresponding to the pop-up menu.

First we need to read in the temperature either from the Edit text box or from the slider. If we read from the slider, the position value is indicated in the variable value. If we read the temperature from the edit text box, this is read directly. We also need to read the minimum and maximum values for the slider. We do this with

```
temp1 = eval(get(handles.input_temp, 'string'))
tmin = get(handles.temp_slider, 'min')
tmax = get(handles.temp_slider, 'max')
```

Note that the value of temperature stored in the edit text box is stored in variable temp1. Next, we need to check if the variable stored in temp1 entered in the edit text box for the temperature is a numerical value, and if it is between the minimum and maximum values for the slider. We do this with

```
if isnumeric(temp1)&...
temp1 >= tmin&...
temp1 <= tmax
```

Then we set the slider bar to the value entered in the edit text box:

```
set(handles.temp_slider, 'value', temp1);
temp = temp1;
```

If the entered temperature is less than the minimum value (set at -500), we set the temperature at -500 degrees and adjust the slider bar accordingly:

```
elseif temp1 < tmin
set(gcbo, 'string', tmin);
set(handles.temp_slider, 'value', tmin);
temp = tmin;
```

We do the same for the case when the temperature is greater than the maximum slider value set at +500 and set the temperature at tmax. This ends the if-elseif-end. Finally, we set the temperature at the variable temp2.

```
elseif temp1 > tmax
set(gcbo, 'string', tmax);
set(handles.temp_slider, 'value', tmax);
temp = tmax;
end
temp2 = temp
```

Now, we add the instructions to read the conversion from the pop-up menu. The variable val indicates which conversion we calculate with:

```
val = get(hObject, 'Value')
switch val
    case 2
        % Celsius to Fahrenheit
        resultt = temp*1.8 + 32;
        set(handles.degrees,'string', 'Fahrenheit')
    case 3
        % Celsius to Kelvin
```

```
        resultt = temp + 273.15;
        set(handles.degrees, 'string', 'Kelvin')
    case 4
        % Fahrenheit to Celsius
        resultt = (temp - 32)*5/9;
        set(handles.degrees, 'string', 'Celsius')
    case 5
        % Fahrenheit to Kelvin
        resultt = (temp - 32)*5/9 + 273.15;
        set(handles.degrees, 'string', 'Kelvin')
    case 6
        % Kelvin to Celsius
        resultt = temp - 273.15;
        set(handles.degrees, 'string', 'Celsius')
    case 7
        % Kelvin to Fahrenheit
        resultt = (temp - 273.15)*1.8 + 32;
        set(handles.degrees, 'string', 'Fahrenheit')
end
```

At this while statement we also give the value to the static text box to the right of the result edit text box. Finally, we write the result to the edit text box result

set(handles.result, 'string', resultt)

We save file with the name Temp_converter_with_slider.m and a run of the GUI is shown in Figure 7.19.

Example 7.3 Control compensation GUI

When we design control systems, an issue of interest is the system stability. Usually, the stability depends upon the compensation criteria. There are several compensation strategies but in this example we only cover three of them. In a control system we start with a plant with a transfer function $G_p(s)$ which may have zeros on the $j\omega$ axis as is the case of motors and integrators. We are usually interested in positioning the poles in such a way that system time domain responses such as overshoot, settling time, rise time, etc., have their characteristics within certain limits. The general feedback control system is shown in Figure 7.20 where $G_C(s)$ is the transfer function of the compensator. The error signal is the difference between the input signal $R(s)$ and the fedback signal $Y(s)$. This signal is the input to the compensator with transfer function $G_{(s)}$. The compensation schemes used in this example are shown in Table 7.3.

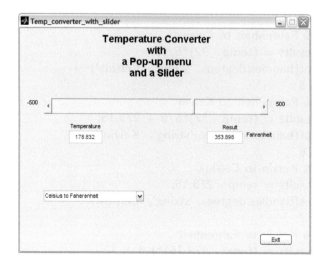

Figure 7.19: Example for the Temperature Converter GUI.

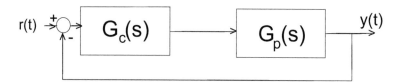

Figure 7.20: Plant with compensator in a loop.

Table 7.3: Compensation schemes

Type of compensation	Transfer function
Proportional (PC)	$G_c(s) = K_c$
Proportional Derivative (PD)	$G_c(s) = K_c + K_D s$
Proportional Integral (PI)	$G_c(s) = K_c + \frac{K_I}{s}$

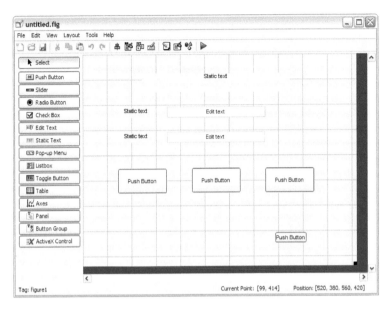

Figure 7.21: Initial compensator GUI layout.

We now design a GUI that realizes these three compensation schemes. The input data to the GUI is the plant coefficients in the format

$$G_c(s) = \frac{\text{Numerator polynomial}}{\text{Denominator polynomial}} = \frac{[a_m a_{m-1}...a_1 a_0]}{[1 \ b_{n-1}...b_1 b_0]} \qquad (7.2)$$

The layout for the GUI is shown in Figure 7.21. In this GUI layout we have resized the components. As in the previous examples we include a button to close the GUI. We change the string for the components so that it looks as in Figure 7.22. Next we change the tags for the push buttons and for the edit text boxes as follows:

Numerator edit text	numerator
Denominator edit text	denominator
PC push button	PC
PD push button	PD
PI push button	PI
Close push button	closeGUI

Now we edit the callbacks. For the push button Close we add the instruction

close(gcbf)

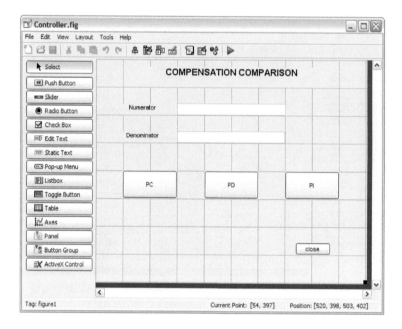

Figure 7.22: Compensator GUI layout.

The callback for the Close push button is then

```
% — Executes on button press in closeGUI.
function closeGUI_Callback(hObject, eventdata, handles)
% hObject handle to closeGUI (see GCBO)
% eventdata reserved - to be defined in a future version of MATLAB
% handles structure with handles and user data (see GUIDATA)
close(gcbf)
```

For the remaining push buttons we add the instruction

PC push button	Proportional()
PD push button	Prop_derivative()
PI push button	Prop_integral()

These instructions mean that when we click on each button, a new GUI opens. The callbacks are then:

For the proportional controller:

```
function PC_Callback(hObject, eventdata, handles)
% hObject handle to PC (see GCBO)
% eventdata reserved - to be defined in a future version of MATLAB
```

```
% handles structure with handles and user data (see GUIDATA)
Proportional()
```

For the proportional derivative controller:

```
function PD_Callback(hObject, eventdata, handles)
% hObject handle to PC (see GCBO)
% eventdata reserved - to be defined in a future version of MATLAB
% handles structure with handles and user data (see GUIDATA)
Prop_derivative()
```

For the proportional integral controller:

```
function PI_Callback(hObject, eventdata, handles)
% hObject handle to PC (see GCBO)
% eventdata reserved - to be defined in a future version of MATLAB
% handles structure with handles and user data (see GUIDATA)
Prop_integral()
```

For the edit text boxes we write

```
nump = ( get( handles.numerator, 'string') )
```

which reads the polynomial numerator. Since this polynomial coefficients are going to be used in another GUI we save them to a mat-file nump.mat with the instruction

```
save nump.mat nump
```

We do the same with the denominator edit text box:

```
denp = (get(handles.denominator, 'string'))
save denp.mat denp
```

The callbacks are then:
For the numerator edit text box:

```
function numerator_Callback(hObject, eventdata, handles)
% hObject handle to denominator (see GCBO)
% eventdata reserved - to be defined in a future version of MATLAB
% handles structure with handles and user data (see GUIDATA)
nump = (get(handles.numerator, 'string'))
save nump.mat nump
```

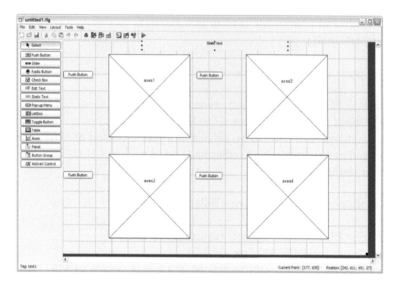

Figure 7.23: Initial GUI layout for compensators.

and for the denominator edit text box:

```
function denominator_Callback(hObject, eventdata, handles)
% hObject handle to denominator (see GCBO)
% eventdata reserved - to be defined in a future version of MATLAB
% handles structure with handles and user data (see GUIDATA)
denp = (get(handles.denominator, 'string'))
save denp.mat denp
```

We now create the individual GUIs for each of the control techniques. In each of the GUIs we read the information from the mat-files and then obtain plots for root-locus, Nichols, and Nyquist diagrams, as well as the step response. The layout for each of the GUIs is the same and only the title text and the feedback equations are different. The layout for the GUIs is shown in Figure 7.23. The strings are modified so that they look as shown in Figure 7.24. Now we modify the tags. For each of the push buttons the tags are modified as: Getrootlocus, GetNicholsplot, GetNyquist, GetNyquistplot, and stepresponse, for Get root locus, Get Nichols, Nyquist diagram, and Get Step. resp, respectively.

The callbacks have to read the information from the mat-files, use it to form the feedback transfer function, calculate the data for the required plot, and plot it in the corresponding set of axes. For the proportional compensator callback the instructions are:

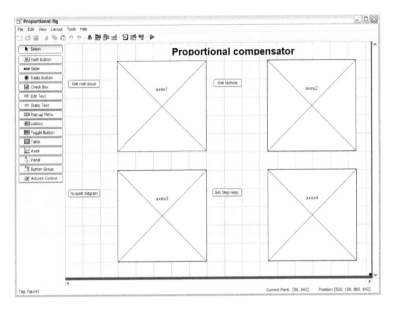

Figure 7.24: GUI with final sizes and strings.

load nump.mat	to read the numerator polynomial from the mat-file nump.dat.
load denp.mat	to read the denominator polynomial from the mat-file denp.dat.
nump1 = str2num(nump)	to convert the string to a numerical format.
denp1 = str2num(denp)	to convert the string to a numerical format.
axes(handles.axes1)	bring the axes to plot.
rlocus(nump1, denp1)	obtain the root locus and plot it.

The final callback for the push button Getrootlocus is

```
function Getrootlocus_Callback(hObject, eventdata, handles)
% hObject handle to Getrootlocus (see GCBO)
% eventdata reserved - to be defined in a future version of MATLAB
% handles structure with handles and user data (see GUIDATA)
load nump.mat
load denp.mat
nump1 = str2num(nump)
denp1 = str2num(denp)
axes(handles.axes1)
rlocus(nump1, denp1)
```

For the Nichols and Nyquist plot we just change the axes where we plot and the function to plot. Thus, we have:

For the Nichols plot:

```
axes(handles.axes2)
nichols(nump1, denp1)
```

and for the Nyquist plot:

```
axes(handles.axes3)
nyquist(nump1, denp1)
```

The complete callbacks for these two push buttons are:

```
function GetNyquistplot_Callback(hObject, eventdata, handles)
load nump.mat
load denp.mat
nump1 = str2num(nump)
denp1 = str2num(denp)
axes(handles.axes3)
nyquist(nump1, denp1)

function GetNicholsplot_Callback(hObject, eventdata, handles)
load nump.mat
load denp.mat
nump1 = str2num(nump)
denp1 = str2num(denp)
axes(handles.axes2)
nichols(nump1, denp1)
```

For the proportional derivative and the proportional integral compensation we have to read in the compensator coefficients and to multiply the compensator transfer function times the numerator of the plant. We do this with

```
nump2 = conv(nump1, [Kd Kp])
```

The complete callback for the Getrootlocus button in the proportional derivative compensation is

```
function Getrootlocus_Callback(hObject, eventdata, handles)
load nump.mat
load denp.mat
nump1 = str2num(nump)
Kp = eval(get(handles.kp, 'string'))
Kd = eval(get(handles.kd, 'string'))
nump2 = conv(nump1, [Kd Kp])
```

```
denp1=str2num(denp);
axes(handles.axes1);
rlocus(nump2, denp1)
```

And for the other push buttons we have the callbacks:

```
function GetNyquistplot_Callback(hObject, eventdata, handles)
load nump.mat
load denp.mat
nump1 = str2num(nump);
denp1 = str2num(denp);
Kp = eval(get(handles.kp, 'string'))
Kd = eval(get(handles.kd, 'string'))
nump2 = conv(nump1, [Kd Kp])
axes(handles.axes4);
nyquist(nump2, denp1);

function GetNicholsplot_Callback(hObject, eventdata, handles)
load nump.mat
load denp.mat
nump1 = str2num(nump);
denp1 = str2num(denp);
Kp = eval(get(handles.kp, 'string'));
Kd = eval(get(handles.kd, 'string'));
nump2 = conv(nump1, [Kd Kp]);
axes(handles.axes5);
nichols(nump2, denp1);

function stepresponse_Callback(hObject, eventdata, handles)
load nump.mat
load denp.mat nump1 = str2num(nump);
denp1 = str2num(denp);
axes(handles.axes6);
Kp = eval(get(handles.kp, 'string'))
Kd = eval(get(handles.kd, 'string'))
nump2 = conv( nump1, [Kd Kp])
plant = tf(nump2, denp1);
system = feedback(plant, [1]);
step(system)
```

For the proportional integral compensation the callbacks are:

```
function Getrootlocus_Callback(hObject, eventdata, handles)
load nump.mat
load denp.mat
```

```
nump1 = str2num(nump)
Kp = eval(get(handles.kp, 'string'));
Ki = eval(get(handles.ki, 'string'));
nump2 = conv(nump1, [1 Ki/Kp]);
denp1 = str2num(denp);
denp1 = [denp1 0];
axes(handles.axes1);
rlocus(nump2, denp1);


function GetNyquistplot_Callback(hObject, eventdata, handles)
load nump.mat
load denp.mat
nump1 = str2num(nump);
denp1 = str2num(denp);
pkp = eval(get(handles.kp, 'string'))
pkd = eval(get(handles.ki, 'string'))
nump2 = conv(nump1, [pkd pkp])
axes(handles.axes4);
nyquist(nump2, denp1);


function GetNicholsplot_Callback(hObject, eventdata, handles)
load nump.mat
load denp.mat
nump1 = str2num(nump);
denp1 = str2num(denp);
pkp = eval(get(handles.kp, 'string'))
pkd = eval(get(handles.ki, 'string'))
nump2 = conv(nump1, [pkd pkp])
axes(handles.axes5);
nichols(nump2, denp1);


function stepresponse_Callback(hObject, eventdata, handles)
load nump.mat
load denp.mat
nump1 = str2num(nump);
denp1 = str2num(denp)
axes(handles.axes6);
pkp = eval(get(handles.kp, 'string'))
pki = eval(get(handles.ki, 'string'))
nump2 = conv(nump1, [1 pki/pkp])
denp2 = [denp1 0]
plant = tf(nump2, denp2);
system = feedback(pkp*plant, [1])
step(system)
```

Figure 7.25: Main GUI for the compensator.

A run for the transfer function

$$G_p(s) = \frac{s+2}{s(s+1)(s+3)} = \frac{s+2}{s^3 + 4s^2 + 3s}$$

produces the main GUI of Figure 7.25 and, for the three types of compensators, the plots shown in Figure 7.26.

7.3 Deployment of GUIs

A GUI can also be deployed for distribution and execution in platforms without a MATLAB license. The procedure is exactly the same described in Chapter 6. We show the procedure with the GUI developed in Example 7.2.

Example 7.4 Deployment of the temperature conversion GUI

The GUI for the temperature conversion GUI is deployed with the following steps:

1. We open the Deployment tool with MATLAB Start →MATLAB Compiler →Deployment Tool. We open a new project with the name TempGUI. prj and we deploy it for a Windows stand-alone application. The files for this GUI are:

Temp_converter_with_slider.m
Temp_converter_with_slider.fig

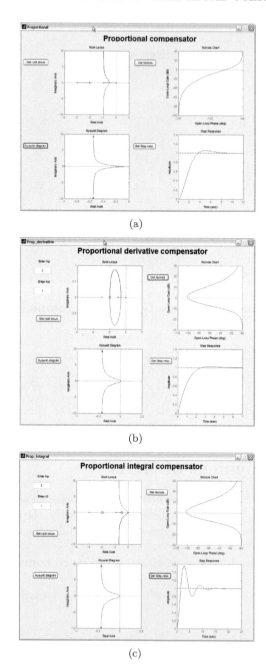

Figure 7.26: Compensator windows. (a) Proportional, (b) Proportional derivative, (c) Proportional integrator.

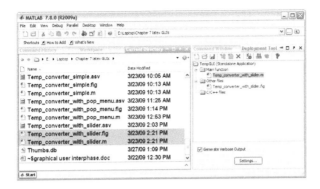

Figure 7.27: File in the directories for deployment.

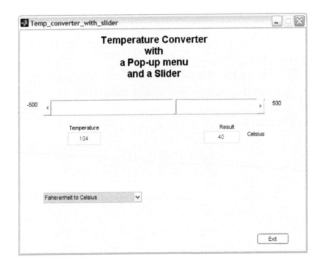

Figure 7.28: GUI for temperature conversion.

We add the file Temp converter_with_slider.m in the Main Function directory and the file Temp converter_with_slider.fig in the directory Other files. This action produces Figure 7.24. Now we click on the Build the Project icon and wait for the tool to finish creating the necessary files. When this task is finished, we click on the Package the project icon and wait for the results. When the packaging finishes we get a new directory TempGUI with two subdirectories: distrib and src. We distribute the directory TempGUI. In the platform without a MATLAB license we follow the procedure outlined in Section 6.8 to install the MATLAB libraries packaged and the executable file. When we run the file TempGUI we obtain Figure 7.28 which is exactly the same as Figure 7.19.

7.4   Concluding Remarks

We have presented the techniques to produce Graphical User Interfaces (GUIs) for MATLAB programs. A GUI provides an interface for input/output data in such a way that the user finds a nice environment to work in. We have presented several examples showing some of the features of a GUI. We have also presented how to deploy a GUI for use in platforms that do not have a MATLAB license.

7.5   Exercises

7.1 Write a GUI that plots the function exp(-x)*sin(2x). Provide a radio button to add a grid or to delete it. Provide an Edit text box to give the initial and final values for x.

7.2 It is desired to have a GUI that takes the Fourier transform of a function. Use an Edit text box to enter the function. Plot the function and the Fourier transform.

7.3 Write a GUI that plots a surface plot of the function $\sin(x, y)$ and add a toolbar to be able to change the observation point.

7.4 Analysis of an RLC circuit produces the differential equation for the current $i(t)$ given by

$$\frac{d^2 i}{dt^2} + \frac{R}{L}\frac{di}{dt} + \frac{1}{LC} i = 0$$

Write a GUI with edit text boxes to enter values for R, L, and C, solve the differential equation, and plot the resulting solution. In particular, try the sets of values: (a) $R=2\ \Omega$, $L=1$ H, $C=0.2$ F, (b) $R=0\ \Omega$, $L=1$ H, $C=0.2$ F, (c) $R=2\sqrt{5}\ \Omega$, $L=1$ H, and $C=0.2$ F, (d) $R=6\ \Omega$, $L=1$ H, $C=0.2$ F.

7.5 To the GUI from Example 7.1 add a mesh plot for the Call Option and Put Option if the stock and strike prices vary between $30.00 and $50.00.

7.6 To the GUI from Example 7.2 add a plot for the temperature conversion required in the range from $0°$ to $100°$.

7.7 Write a GUI where the user enters a square matrix, its dimensions and elements. It has to display the determinant, trace, inverse, and eigenvectors and eigenvalues.

7.8 A GUI has as input two functions $f(x)$ and $g(x)$. The GUI must differentiate and integrate the functions and calculate them at a user-given point.

7.9 To the previous example add the composition $f(g(x))$of the two functions and evaluate it.

7.10 Write a GUI with Edit text boxes to enter two two-dimensional vectors. Obtain the cross-product and plot the vector in a three-dimensional plot.

7.6 References

[1] P. Marchand and O. T. Thomas, Graphics and GUIs with MATLAB, 3rd Ed., Chapman & Hall/CRC, Boca raton, FL, 2003.

[2] D. C. Hanselman, Mastering MATLAB 7, Prentice Hall, Inc., Piscataway, NJ, 2004.

[3] MATLAB Creating Graphical User Interfaces, The MathWorks, Natick, MA, 2009.

[4] H. Moore, MATLAB for Engineers, 2nd Edition, Prentice Hall, Inc., Piscataway, NJ, 2008.

Chapter 8

Simulink

Simulink is a package that is useful to model, simulate, and analyze systems, either continuous time or discrete time. They may also be either linear or non-linear, and the systems modelled in Simulink may be a combination of them. Simulink has a graphical interface that makes it very easy to build models and then simulate them. It also has a set of components grouped in libraries in different topics of engineering. Many of the libraries can be used in different topics, such as physics, finance, mechanical engineering, hydraulics, etc. Many of these libraries can be found in toolboxes which are available separately.

8.1 The Simulink Environment

To start working with Simulink we need to be working first with MATLAB. In the MATLAB main window we can either write simulink in the Command window or click on the Simulink icon in the MATLAB main toolbar. The Simulink icon is shown in Figure 8.1. This action opens the Simulink main window, called Simulink Library Browser, shown in Figure 8.2. To the right we can see all the libraries and toolboxes available for modelling and simulation. In the context of Simulink, the toolboxes are also called blocksets. Figure 8.3 shows the blocks available for the library Continuous which is available within the Simulink library. The libraries, toolboxes, and blocksets available for modelling and simulation depend upon which ones have been bought and the user needs. In this chapter our approach is on trying to get acquainted with Simulink details and we leave the details of particular libraries for the interested reader to consult the references.

Figure 8.1: Icon for Simulink.

Figure 8.2: Simulink main window.

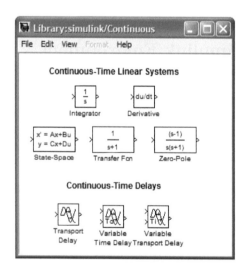

Figure 8.3: Blocks available for the **Continuous** set within the Simulink library.

8.1.1 A Basic Example

To create a new model for a system, in the File menu we select New (alternatively we can click on the New icon). This opens a model window. Then, we open the library we need and drag the required blocks to our model window. The blocks are connected together with lines. Finally, we add source signals to excite the system model and sinks to see the outputs. With a simple low pass transfer function we show the procedure.

Example 8.1 Simple model for a low pass filter

Let us consider the second-order transfer function

$$N(s) = \frac{1}{s^2 + s + 1}$$

This transfer function appears frequently in engineering and mathematics. For example, in electronics it represents a control system transfer function, in physics in a damped string or pendulum, in finance a cash flow, and in many more applications.

This transfer function is excited by a step input and we wish to observe the output. We open a new model window, shown in Figure 8.4(a), and from the Continuous library we drag to it a Transfer function block, from the Sources library we drag a Step signal, from the Sinks library we drag a Scope, and from the Signal Routing library we drag a Mux block.

The model window with the blocks just dragged is shown in Figure 8.4(b). Note that the transfer function block has two small > symbols, one to the right and one to the left of the block, while the Scope and the step input have only one > symbol in the left side and in the right side, respectively. The > symbols are either inputs or outputs to the blocks. If the > symbol points outwards it is an output and if it points inwards it is an input.

Now, we need to connect the blocks. There are two ways to connect two blocks. The easiest way is to select a block, hold down the CONTROL key and left-click on the destination block. If necessary, Simulink routes the line around any intervening blocks. If there are several inputs and outputs between the two blocks, Simulink draws as many lines as possible between the two blocks. The other way to connect two blocks is by placing the pointer in the output of a block, the pointer changes to a crosshair, left-click and drag it to the input of the desired block.

Using any of the techniques just described we first connect a line from the Step to the input of the transfer function block. Then, we connect a line from the output of the transfer function block to the one of the inputs of the Mux block. Finally, we connect the output from the Mux to the input of the Scope. The model now is shown in Figure 8.4(c). To the remaining Mux input we connect the Step block. To do this we place the pointer on the line that goes from the Step to the Transfer function block, right click and when

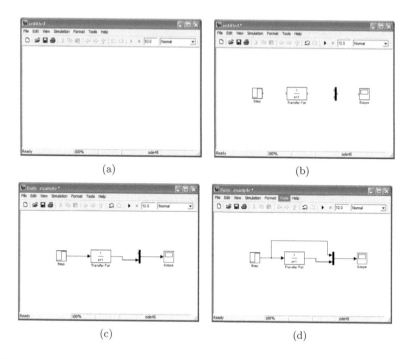

Figure 8.4: Steps for the creation of a new model. (a) New model window, (b) Blocks necessary for the model, (c) Lines joining the blocks, (d) Final model.

the crosshair appears we drag it to the unconnected Mux input. This step concludes the connections among the model blocks. The model now is shown in Figure 8.4(d). Now we need to enter the correct transfer function in the Transfer function block. We double-click on the block to open the Properties window where we change the coefficients in the denominator from [1 1] to [1 1 1] as shown in Figure 8.5. In the Step properties window we also make Step time= 0. The next step is the simulation of the model. We just click on the Run icon in the model window. When the simulation ends, we click on the Scope block and the window for the Scope signals opens showing the signal output for the transfer function. Figure 8.6 is produced after clicking on the Autoscale icon in the Scope toolbar.

The example above shows the process we have to follow to model and simulate a system. In the following sections we introduce other characteristics of Simulink.

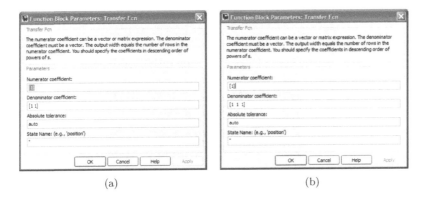

Figure 8.5: **Properties for the transfer function block, (a) Default values for the denominator polynomial, (b) Desired values for the denominator polynomial.**

8.2 Model Edition

There are a number of options that we can make on blocks and lines in a model. We describe them below.

Selecting a block.	To select a block we just single click on it.
Label changing.	Click on the block label. After an editing cursor appears, just type the desired label.
Change label location.	Select the block and from the **Format** menu select **Flip Name**.
Block resizing.	After selecting the block, click on a handle and resize as desired.
Adding a shadow.	Select the block and from the **Format** menu select **Show Drop Shadow**.
Rotating a block.	Select the block and from the **Format** menu select either **Rotate** or **Flip Block**.

For lines we can make the following actions:

Moving a line.	Select the line and drag it to the desired place.
Moving a vertex.	Click on the vertex and drag it to the desired place.
Deleting a line.	Select the line and press the Delete key.
Labeling a line.	Double-click the line. When the editor appears, write the label.
Moving a line label.	Select the line label and drag it to the desired place.
Editing a line label.	Select the label and edit it.

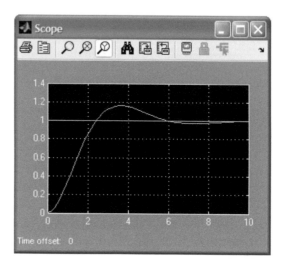

Figure 8.6: Input and output signals in Scope.

Additionally we can **annotate** the model. To make an annotation, double-click on the place we wish to make the annotation and when the editor appears enter the annotation. Click away from the annotation to finish the process.

8.3   The Scope

In the Scope window of Figure 8.6 we have a toolbar that allows us to operate on the plotted signals. Figure 8.7 shows the tasks that each icon performs. The most important ones are the Autoscale, Save current axis settings, and Restore saved axis settings. With the Autoscale icon we use the full scale of the Scope. This icon was used to generate the final plots shown in Figure 8.6. With the Save Current axis button we can store the current x and y axis settings so we can apply them to the next simulation. Another important icon in the Scope toolbar is the Floating Scope one. This icon floats and unfloats the Scope. A floating Scope is used to watch the signals during a simulation.

8.4   Continuous and Discrete Systems

A continuous model is composed of continuous blocks. These blocks are available in the Continuous library. Blocks from other libraries can also be used in continuous systems with the exception of the blocks in the Discrete and the Logic and Bit Operations libraries. In continuous systems the time varies continuously and in the frequency domain they use the complex frequency (Laplace transform) variable s. A discrete model uses blocks from the Dis-

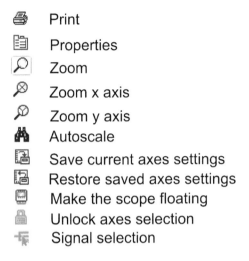

🖨	Print
📑	Properties
🔍	Zoom
⊗	Zoom x axis
⊘	Zoom y axis
🔍	Autoscale
	Save current axes settings
	Restore saved axes settings
	Make the scope floating
	Unlock axes selection
	Signal selection

Figure 8.7: Icons in Scope.

crete library as well as blocks from any other library with exception of blocks from the Continuous library. In discrete systems the time changes in steps T known as the Sample time and it is not continuous. A system that has blocks from Continuous and Discrete libraries is called a Hybrid or Mixed-mode system. In these models the time can be either continuous or discrete. We present examples of these three kinds of systems.

Example 8.2 A continuous system

We already have a continuous model in Example 8.1. Another example is the modelling of the behavior of a pendulum. The pendulum differential equation for the case of a small swing is

$$mL\ddot{\theta} + BL\dot{\theta} + W\theta = 0.04\delta(t) \qquad (8.1)$$

where θ is the angular position, $\dot{\theta}$ is the angular velocity, m is the pendulous mass, $L=$ 0.6 m is the length of the pendulum arm, $B=$0.08 Kg/m/s is the damping coefficient, and $W=$ $mg=$ 2 Kg is the weight. We are assuming that $g=$ 9.8 m/s² and $m=$ 0.2041 Kg. This differential equation can be written as a set of linear differential equations. First define the state variables $x_1(t) = \theta$ and $x_2(t) = \dot{\theta}$. Then, we can write the differential equation as a system of first-order differential equations:

$$\dot{x}_1 = x_2$$

$$\dot{x}_2 = -16.33x_1 - 0.3920x_2 - 0.04\delta(t)$$

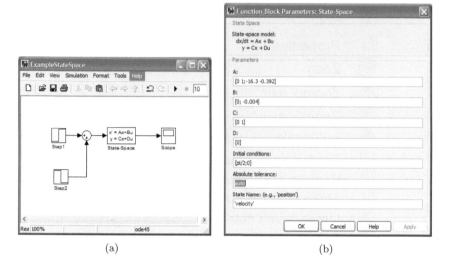

(a)   (b)

Figure 8.8: Continuous model, (a) Complete continuous model, (b) State-space matrices parameters.

This system of equations can be written in matrix form as

$$\begin{bmatrix} \dot{x}_1 \\ \dot{x}_2 \end{bmatrix} = \begin{bmatrix} 0 & 1 \\ -16.33 & -0.392 \end{bmatrix} \begin{bmatrix} x_1 \\ x_2 \end{bmatrix} + \begin{bmatrix} 0 \\ -0.04\delta(t) \end{bmatrix}$$

$$\begin{bmatrix} y_1 \\ y_2 \end{bmatrix} = \begin{bmatrix} 0 & 1 \end{bmatrix} \begin{bmatrix} x_1 \\ x_2 \end{bmatrix}$$

The initial conditions are

$$\begin{bmatrix} x_1(0) \\ x_2(0) \end{bmatrix} = \begin{bmatrix} \pi/2 \\ 0 \end{bmatrix}$$

This system is modelled in Simulink with the **State-space** block from the Continuous library. The input to this block is an impulse and the output is seen in a Scope. The complete model is shown in Figure 8.8(a). For the **State-space** block we enter the system matrix A and the vectors B, C, and D together with the initial conditions as shown in Figure 8.8(b). The impulse is created with two step signals subtracting the second from the first but the second one delayed 0.01 sec from the first one that has the step at 0 sec. After running the model we see the response as shown in Figure 8.9(a). The output signal is not as good as we might expect. This is due to the fact that in the simulation

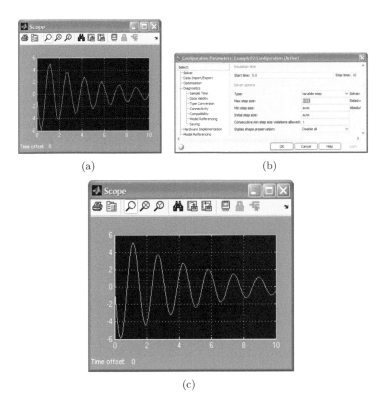

(a)   (b)

(c)

Figure 8.9: Output signal, (a) Poor output due to poor sampling, (b) Changing the maximum time step, (c) Improved output signal.

the step is set to Auto as can be seen in Figure 8.9(b) which is opened from the Simulation menu with Configuration Parameters. Changing the maximum step size to 0.01, clicking on the OK button and running the simulation again we get the output shown in Figure 8.9(c) which is of a better quality. From the output we conclude that the pendulum decreases the amplitude of the oscillations and that eventually it will stop.

Example 8.3 A discrete system

A discrete system has blocks from the Discrete library. Discrete models appear in digital signal processing (DSP), at discrete-time signal processing such as in finance, switched-capacitor filters in electronics, finite difference methods in engineering, to mention a few. The variable in discrete models is the z-transform variable z and the time is in discrete steps. As an example, let us consider the model shown in Figure 8.10(a). It is a digital filter and the basic block is the unit delay block represented by $1/z$. This model realizes the transfer function

$$H(z) = \frac{z^2 - 1}{z^2 - 0.75z - 0.794}$$

The response is shown in Figure 8.10(b) and we see that it is a discrete time waveform. The input signal is a unity frequency signal with an amplitude of 1 and a sample time of 0.5 s.

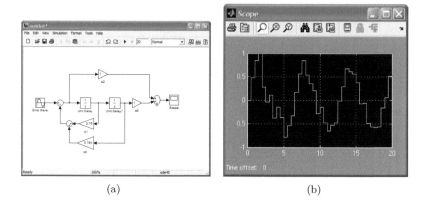

(a)   (b)

Figure 8.10:   Discrete-time model, (a) Complete discrete-time model,
(b) Discrete-time output signal.

Example 8.4 A mixed-mode model

For an example of a mixed-mode system let us consider the cascade of a continuous transfer function block and of a discrete-time one. The system excitation is a step signal that starts at 0 sec. The model is shown in Figure 8.11(a) and the response in Figure 8.11(b). We readily see that the response is a discrete-time signal.

8.5   Subsystems

Similar to the case of functions in MATLAB and subroutines in other programming languages, portions of a model can be grouped in a model for reuse and customization. This is especially useful when the model has grown significatively. A subsystem is a hierarchical grouping of blocks encapsulated by a

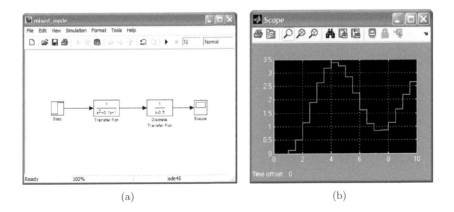

(a)   (b)

Figure 8.11: Mixed-mode model, (a) Complete mixed-mode model, (b) Output signal.

single subsystem. A subsystem can have many blocks that could make hard to understand the complete model. A subsystem can be reusable, thus if the same subsystem function is needed again, we can repeat the subsystem within the model.

There are two ways to create a subsystem. In the first method we only need to select a group of blocks and then from the Edit menu select Create Subsystem (or Control+G). In the second method, we start with an empty Subsystem block available in the library Ports & Subsystems and we place the blocks and lines needed for the subsystem. We show these procedures with examples.

Example 8.5 Creation of a subsystem from a complete model

Let us consider the model shown in Figure 8.12(a). This is a state variable filter. The first step to create a subsystem is to select the parts of the system that we wish to have in the subsystem. For our example, these blocks and lines for the subsystem are shown in Figure 8.12(b). Next, from the Edit menu [see Figure 8.12(c)] we select Create Subsystem. This action creates the subsystem and the model window looks like it is seen in Figure 8.12(d) where we see that now the components of the model selected before are now inside the subsystem. If we double-click on the subsystem block, we see the components with input and output ports added by Simulink as can be seen in Figure 8.13.

To reverse the subsystem, from the Edit menu, we select Undo Create Subsystem. The subsystem block will be replaced by the original set of blocks and lines. Any changes made to the subsystem after its creation will be removed. However, we can recover all the changes made to the subsystem

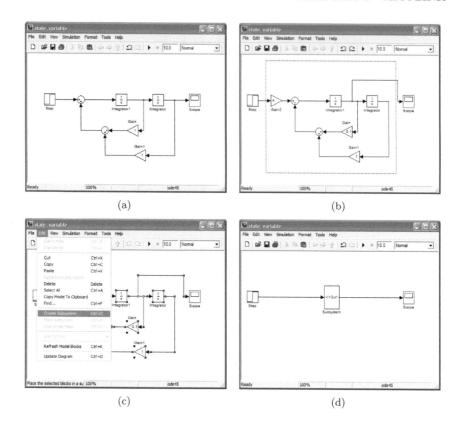

Figure 8.12: Subsystem from an existing model, (a) Original model, (b) Selection of the blocks and lines for the subsystem, (c) Path to create the subsystem, (d) Main model window.

with Edit → Redo Create Subsystem.

Example 8.6 Creating a subsystem using the Subsystem block

To create a subsystem in this way, in a model window we drag a Subsystem block from the Ports & Subsystems library, as shown in Figure 8.14(a). We double-click on the block and see that there are only input and output ports and a line connecting them as we see in Figure 8.14(b). We edit the subsystem model as appropriate. In this example we only add an integrator, as shown in Figure 8.14(c). We close the subsystem model window and return to the main model window where we add a Pulse generator from the Sources library and a Scope from the Sinks library as can be seen in Figure 8.14(d). Finally, we run the simulation by clicking on the Run icon or by Control+T.

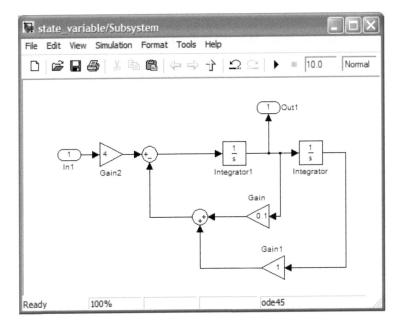

Figure 8.13: State variable filter encapsulated in a subsystem.

8.5.1 Masking Subsystems

Masking enables us to use a subsystem as an independent block thus extending the concept of abstraction. A masked block can have a custom icon and a dialog box of its own where we can enter parameters as if it were a Simulink block. Each and every block available in the Simulink variables is masked. The procedure to mask a subsystem consists of the following steps:

1. We first build a subsystem.
2. In the main model window, we select the subsystem block.
3. Select the subsystem block and select Edit → Mask Subsystem. The Mask Editor Subsystem window is open.
4. In the Mask Editor Subsystem window, set up the information in the tabs.

We show this procedure with an example.

Example 8.7 Masking a subsystem block

Let us consider again the state variable filter subsystem. Now we wish to change the values of the gain blocks to A1 and A0. This can be done by double-clicking on each of the gain blocks and changing the Gain parameters from their values to A1 and A0 as shown in Figure 8.15(a) for the Gain block A1. The final subsystem is shown in Figure 8.15(b). Now, the gain parameters

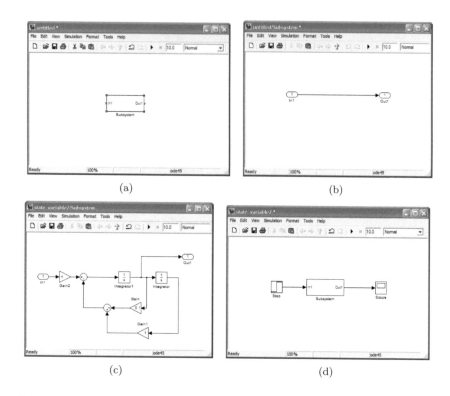

Figure 8.14: **Subsystem created from a subsystem block, (a) Empty subsystem block, (b) Initial components in the subsystem, (c) Final subsystem, (d) main model window.**

of the model are the values of the Gain blocks whose values are given by the coefficients of the transfer function. We now mask this subsystem in such a way that the values of the gain blocks are given in a dialog box. In the main model window we select the subsystem block and then Edit →Mask Subsystem. This opens the dialog window for the Mask Editor for a subsystem which is shown in Figure 8.16. The mask editor has four tabs: Icon, Parameters, Initialization, and Documentation. We now describe each of these tabs' tasks.

8.5.2   Icon Tab

The Icon dialog window allows us to create an icon for the subsystem block. As can be seen in Figure 8.16(a), there is a window where we enter the commands that draw the icon on the subsystem block. Among the options we can execute for our block's icon we have:

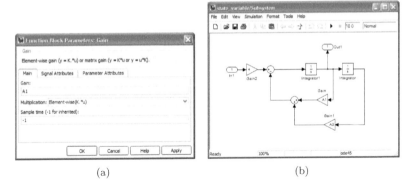

(a)   (b)

Figure 8.15: Masking a subsystem block, (a) Gain parameter changed to **A1**, (b) Subsystem with Gain parameters changed to **A0** and **A1**.

Plot a curve	Use MATLAB plotting instructions.
Display an image	The image must be in the model's folder.
Name the I/O ports	Use the instruction port_label('input').. or 'output', port number,'label').
Transparency	Set the Transparency pulldown menu in the Icon options column.
Rotation	Set the Rotation pulldown menu in the Icon options column.
Change of units	Set the Units pulldown menu in the Icon options column.
Change the color	Use the color code from MATLAB. Place before an instruction.
Display a transfer function	Use poly(numerator, denominator).
Plot poles and zeros	Use droots(zeros vector, poles vector, constant).
Print a text	Use text('text to write').
Print a formatted text	Use fprintf as in MATLAB.

As examples the reader could try the following commands:

```
port_label('input', 1,'Filter in')
color('blue')
text(0.5, 0.5, 'State Var')
plot([1,0], [0.3, 0.8])
```

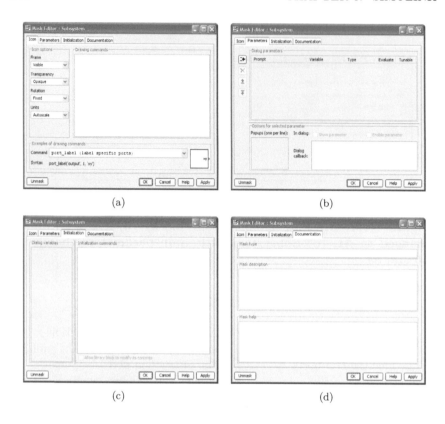

Figure 8.16: **Mask Editor** pane, (a) **Icon** tab, (b) **Parameters** tab, (c) **Initialization** tab, (d) **Documentation** tab.

8.5.3   Parameters Dialog Window

The **Parameters** pane is shown in Figure 8.16(b). In this pane we enter the parameters for the subsystem which in our example are the coefficients A0 and A1. To do this we click on the top button at the left to open a new window shown in Figure 8.16(a) to add a new parameter to the now empty list. At the **Prompt** we add **Coefficient A0** at the **Coefficient** box and at the **Variable** box we add A0. We repeat for coefficient A1. The final window is shown in Figure 8.17(a). These two parameters will appear at the dialog window for the block when we double-click on it once the masking procedure has been finished. After we press the button **Apply** we double-click on the subsystem block and the dialog window of Figure 8.17(b) will prompt for the values of the coefficients A0 and A1.

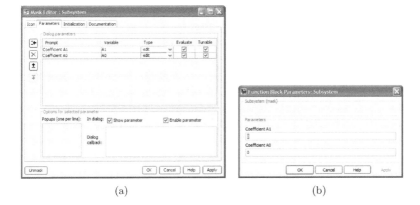

Figure 8.17: Properties pane, (a) Entering coefficients, (b) Parameters dialog window.

8.5.4 Initialization Tab

In this window we can enter MATLAB instructions that allow an initialization of the subsystem block. The block is initialized when:

The model is loaded
The simulation is started or the block is updated
The block is rotated
The block's icon is redrawn

We can additionally enter MATLAB commands in the Initialization Commands box to initialize the masked subsystem.

8.5.5 The Documentation Tab

This tab allows us to define or modify the type, the description and the help text for the masked subsystem block. For example, for our subsystem we can enter the following text in the Mask type dialog box:

State variable filter model.

In the Mask description box we can enter:

This is a realization in state-variable form of a second order transfer function. It is implemented by two integrators, two gain blocks, and two adders.

Finally in the Mask help we can enter:

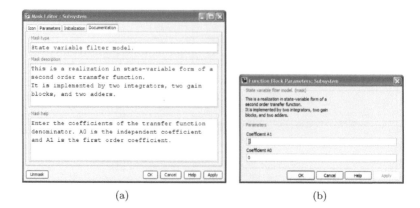

(a)   (b)

Figure 8.18: Documentation tab, (a) Information entered, (b) Block Parameters dialog window.

Enter the coefficients of the transfer function denominator. A0 is the independent coefficient and A1 is the first-order coefficient.

The Documentation window is shown in Figure 8.18(a). After clicking on the Apply button, we double-click on the masked subsystem block and obtain the Block Parameters dialog window for the block shown in Figure 8.18(b).

8.6   Model Linearization with the Control Toolbox

The Simulink Control toolbox has a tool that allows to obtain several model responses by linearizing the model. We show this tool with an example.

Example 8.8 Control of a satellite attitude

A satellite is characterized by a plant transfer function

$$H(s) = \frac{1}{s^2}$$

The state variables are the angular position θ and the angular velocity $\dot{\theta}$. A system with state-variable feedback is shown in Figure 8.19. For values of $K_1=4$ and $K_2=32$, it is desired to find the step and impulse responses as well as a Bode plot. We can do this using the Control toolbox Linearization tool.

To do this, from the menu Tools select Control Design→ Linear Analysis as shown in Figure 8.20. This opens the window of Figure 8.21 for the Control Estimation and Tools Manager shown in Figure 8.20. Here at the lower right

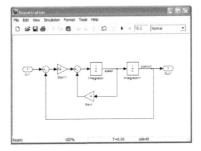

Figure 8.19: State variable feedback for satellite attitude control.

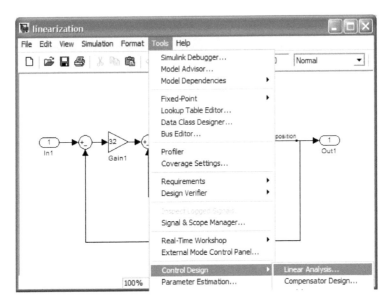

Figure 8.20: Path for Linearization.

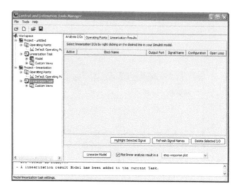

Figure 8.21: Window for Linear analysis.

corner we select the required response from the pulldown menu. The available options are:

Step response plot
Impulse response plot
Bode response plot
Bode magnitude plot
Nyquist plot
Nichols plot
Singular value plot
Pole-zero map
I/O pole zero map
Linear simulation
Initial condition response

In our example we only require to plot the impulse and step responses and the Bode plot. Selecting the step response plot and clicking on the Linearize model button we get the step response plot shown in Figure 8.22(a). We also request the impulse response and the Bode plots.

Now we change the value of K_2 to 8 (its previous value was 4) and run again the same linearization. The plots obtained now are shown in Figure 8.23. We see there that the responses have improved because the overshoot has been reduced. The impulse response is smoother and the Bode plot magnitude is flat.

8.7 Examples

We present examples of different systems that can be modelled in Simulink. The first example is a digital circuit, this example shows another Simulink strength that has been overlooked for some time. The second example is an

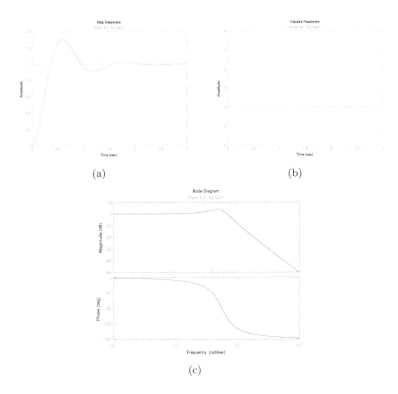

Figure 8.22: Responses for satellite attitude for $K_1=4$. (a) Step response, (b) Impulse response, (c) Bode plot.

application in finance which can be modelled as a discrete-time model.

Example 8.9 Modelling of a digital sequential circuit

Let us consider the counter with unused states defined by the state diagram of Figure 8.24(a). A Simulink model for this counter is shown in Figure 8.24(b). We are using the JK flip-flops that are available in the library Simulink Extras. The JK flip-flops can be initialized to an initial state of 1 or 0. In our circuit we initialize the flip-flops to 1. This can be done by double-clicking on the flip-flops and entering a unity value for the initial. The JK flip-flops are negative edge triggered. The digital clock is set to a Period= 0.5. We now run the simulation and see the output signals at the Scope shown in Figure 8.25. There we can see that the initial state for bits A, B, and C is 111. (The least significant bit is A.) When the first negative edge clock happens, the counter changes state to 000. The reader can observe that after each negative edge the output changes to another state according to the state diagram of Figure 8.24(a).

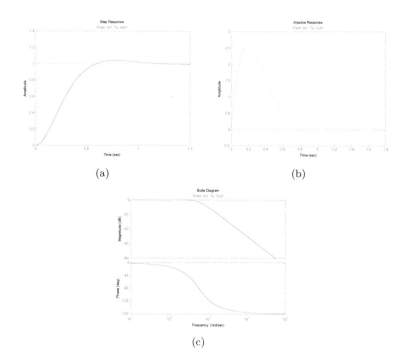

Figure 8.23: Responses for satellite attitude for K_1=8. (a) Step response, (b) Impulse response, (c) Bode plot.

Example 8.10 Savings for retirement

A 25-year-old man is planning to retire when he reaches the age of 65 years. He begins a savings plan starting with an initial deposit of $10,000. He wishes to know how much he is going to have at the end of his productive life if he saves $200 each month at an annual interest rate of 6%. The data for this problem is: initial condition = 10,000, monthly interest rate = 0.5, months in plan = 40×12 = 480. The problem can be modelled by the difference equation

$$\mathrm{sum}(nT) = \mathrm{sum}[(n-1)T] * \mathrm{interest\_rate} + \mathrm{monthly\_deposit}$$

A Simulink model that realizes this equation is shown in Figure 8.26. Here, we have a **constant** block as input with a value of 200 which is the monthly deposit. In addition, we have a time delay which represents the month that passes between deposits. The feedback amplifier represents the increase in value of the deposit due to the monthly interest. This is added to the monthly deposit to have the new sum. Finally, we are displaying the sum in the **Scope**, in a **Display** block and in a **To Workspace** block (both blocks are available in the **Sinks** library) to send the numerical data to MATLAB. The initial condition in the **Delay** block is set to 10,000 to account for the initial deposit.

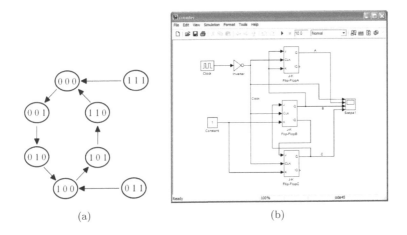

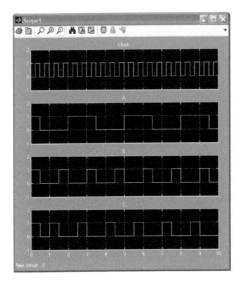

Figure 8.24: Digital counter with unused states. (a) State diagram,
(b) Simulink model of the digital counter.

Figure 8.25: Digital counter output signals.

The Gain block value is 1.005 which is the original amount plus the monthly
interest of 0.5%. The simulation is run for 480 seconds to account for the 480
months in the plan. We now run the simulation and observe that the display
block shows that the final amount saved is $507,872.68. The Scope output of
Figure 8.26(b) shows the way the savings increase. Note that if the money
were not invested, the savings at the end of the plan would have only been
$106,000. If we now go to MATLAB we see that in the Workspace window
we have two variables, simout and tout. tout is a vector of length 480 which
is the number of months. simout is an array which has three elements: time,
signals, and blockName. Of these elements, signals has three components one
of them being values. If we use the icon plot on values we get Figure 8.26(c)
which is identical to the Scope output.

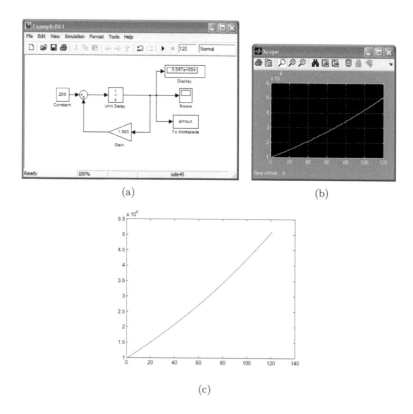

(a)

(b)

(c)

**Figure 8.26: Model for retirement savings. (a) Output from Scope,
(b) Output from Simulink.**

8.8 Concluding Remarks

This chapter cover a very useful and important application of MATLAB. This application is called Simulink and it is block-design oriented. The examples presented show that it is very easy to use. In the same way as MATLAB, it can be enriched with toolboxes that are especially designed for certain areas of science, and engineering. An important topic presented is the creation of subsystems and the masking of these subsystems. Finally, some examples show some of the capabilities of Simulink.

8.9 Exercises

8.1 A mass-spring-dashpot system is described by the equation

$$m\ddot{x} + c\dot{x} + kx = F$$

where m is the mass, F is the acting force on the mass, and k is the spring constant. Model the system on Simulink and plot the results on a scope. Use k=10 N/m, c= 0.1 N/m/kg, F= 5 N, and m= 1 kg.

8.2 The temperature conversion equation to change degrees Celsius to Fahrenheit is

$$F = 1.8C + 32$$

Model this equation in Simulink and send the result to MATLAB.

8.3 A 20-kg chair is hanging attached to a spring and allowed to oscillate. When the chair is empty the chair oscillates with a 1.5-s period. Model the motion in Simulink. Then, at t= 8 s a 60-kg person seats on the chair. Model the oscillation starting at t=0 s and ending at t=15 s for the complete system.

8.4 Make a subsystem for the pendulum of Example 8.2. Then mask the subsystem so the user can enter the parameters from dialog windows and repeat Example 8.1.

8.5 The equations of motion of a ball thrown at an angle θ are

$$m\ddot{x} = 0$$

$$m\ddot{y} = -mg$$

Model these equations in Simulink and plot the path for a ball with m= 0.1 kg.

8.6 A system has the transfer function

$$G(s) = \left(\frac{1}{s+1}\right)^2$$

Obtain the input and step responses for the linearized system.

8.10   References

[1] J.B. Dabney and T.L. Hartman, Mastering Simulink, Prentice Hall, Inc., Upper Saddle River, NJ, 2003.
[2] S. Karris, Introduction to Simulink with Engineering Applications, Second Edition, Orchard Publications, 2008.
[3] H. Klee, Simulation of Dynamic Systems with MATLAB and Simulink, CRC Press, Boca Raton, FL, 2008.
[4] M. Nuruzzaman, Modeling and Simulation in SIMULINK for Engineers and Scientists, AuthorHouse, Bloomington, IN, 2005.
[5] Simulink Users Guide, The MathWorks, Inc., Natick, MA, 2009.

Chapter 9

MATLAB Applications in Engineering

In this chapter we present some applications to fields in engineering. Electronic, civil, mechanical, chemical and food engineering applications are covered. Of course, we do not pretend to be exhaustive because that would be impossible to do. However, we present some representative examples in those areas. The examples presented include applications of some of the toolboxes available from The MathWorks, Inc., while other examples have scripts written specifically for them. Among the toolboxes used we have the Optimization, Symbolic Math, Control, and Signal Processing toolboxes.

9.1 Applications in Signals and Systems

Frequently, engineering systems are modelled either by a differential equation or by a transfer function. Any of these descriptions provide a great deal of information about the system. For example, we can obtain information about the stability of the system, its step or impulse response, and its frequency response. This section uses the Signal Processing toolbox.

Example 9.1 Second-order transfer function behavior

Bode plots are plots of magnitude in dB and phase in degrees of the transfer function N(s). Let us consider the second-order transfer function

$$N(s) = \frac{\omega_n^2}{s^2 + 2\omega_n\xi s + \omega_n^2} \qquad (9.1)$$

where ω_n and ξ are the natural frequency and damping coefficient, respectively. This transfer function has applications in control and analog filtering. We wish to know the response of $N(s)$ when $\omega_n = 2$ and $\xi = 0.1$. For these

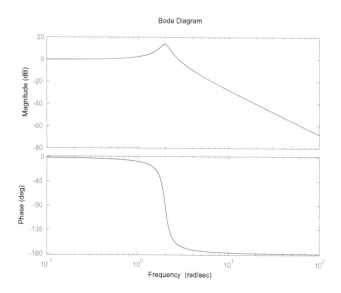

Figure 9.1: Bode plots.

values, $N(s)$ is

$$N(s) = \frac{4}{s^2 + 40s + 4}$$

We can obtain a plot of its magnitude and phase with the instruction **bode**. This instruction produces plots of magnitude in dB and phase in degrees. The instruction **bode** requires to give the numerator and denominator polynomials. For our function we have

$$\text{num} = [4]$$

$$\text{den}[1\ 40\ 4]$$

Then, the instruction bode is

$$\text{bode(num, den)}$$

This will produce the Bode plots shown in Figure 9.1. Note that the function N(s) corresponds to a low-pass filter. Also note that the phase is negative. It has a minimum value of $-180°$. This is so because the transfer function has two poles and each pole contributes a maximum of -90° to the total phase. Now, let us consider the plot for the function

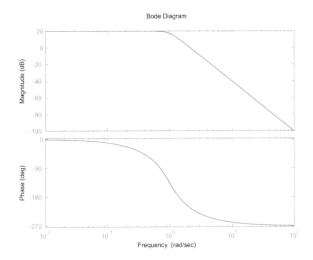

Figure 9.2: Bode plots for a third-order function.

$$N(s) = \frac{10}{s^3 + 2s^2 + 2s+}$$

The Bode plots for this function are shown in Figure 9.2. In this case the phase plot minimum value is $-270°$ because the transfer function has three poles.

If we now wish to see the behavior of the second-order function for different values of ξ we can use the following script that uses a for instruction to increment ξ values,

```
% This is file Example9_1.m
clc
clear
close all
num = [1];
zeta = [ 0.1 0.2 0.3 0.4 ];
for k = 1:4;
       den = [1 2*zeta( k ) 1];
       bode(num, den);
       hold on
end
hold off
```

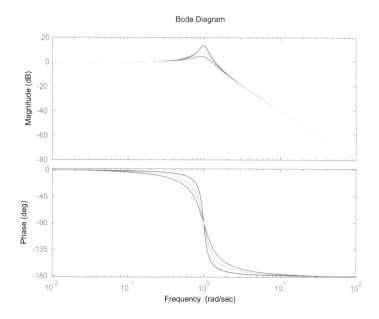

Figure 9.3: Effect of changing ξ in a second-order transfer function.

The result is shown in Figure 9.3. There we see that the effect of changing the parameter ξ affects the peak in the magnitude and the phase around the cutoff frequency.

Example 9.2 Time domain response of a second-order function

The unit-impulse and the unit-step response of the second-order transfer function can be obtained with the instructions impulse and step which have the format

$$\text{step(numerator, denominator)}$$

For the second-order transfer function

$$N(s) = \frac{16}{s^2 + 2s + 16}$$

we have then that the impulse and step responses can be obtained with

```
num = [16];
den = [1 2 16];
impulse(num, den)
```

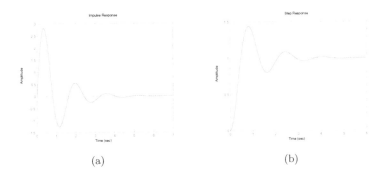

(a)　　　　　　　　　　　　　　　　　(b)

Figure 9.4: Unit and step responses, (a) Unit-impulse response, and (b) Unit-step response.

```
figure
step(num, den)
```

which produces the plots shown in Figure 9.4. We now wish to know the unit-step response of the second-order transfer function when ξ changes. For example, for the values of ξ 0.1, 0.2, 0.3, and 0.4, we can plot the step response with the following script:

```
num = [1];
for zeta = 0.1: 0.1: 0.4;
    den = [1 2*zeta 1];
    t = 0: 0.1: 19.9;
    step(num, den, t)
    hold on
end
xlabel('time')
ylabel('step response')
legend('zeta = 0.1','zeta = 0.2','zeta = 0.3','zeta = 0.4')
```

We can readily see that the overshoot is larger for smaller values of ξ. The same information available in Figure 9.5 can be plotted in a three-dimensional plot with:

```
k = 1;
for zeta = 0.1: 0.1: 0.4
    den = [1 2*zeta 1];
    t = 0: 0.1: 19.9;
    y(:, k) = step(num, den, t)
    k = k+1;
```

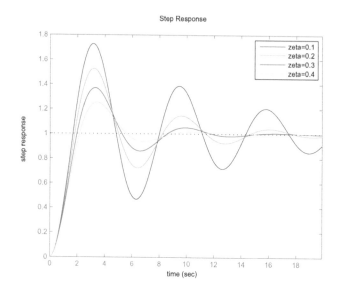

Figure 9.5: Step response for different values of ξ.

```
end
mesh(y)
xlabel('zeta')
ylabel('time')
zlabel('Step response')
```

The plot shown in Figure 9.6 gives an idea of how the step response changes with the parameter ξ.

Example 9.3 Laplace Transforms

The Laplace transform of a piecewise continuous function $f(t)$ for $t \geq 0$ is defined by

$$\mathcal{L}\{f(t)\} = F(s) = \int_0^\infty f(t)e^{-st}dt \tag{9.2}$$

where s is the complex frequency variable $s = \sigma + j\omega$. On the other hand, the inverse transform is given by

$$f(t) = \frac{1}{2\pi j} \int_{c-j\infty}^{c+j\infty} F(s)e^{st}ds \tag{9.3}$$

where $c > 0$. MATLAB allows us to obtain the direct and inverse Laplace

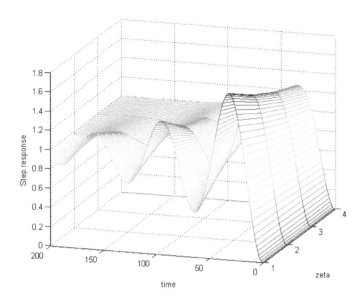

Figure 9.6: Mesh plot of step response for different values of ξ.

transforms of a function with the instructions laplace and ilaplace, respectively. For example, for the functions

$$f_1(t) = t^3 e^{-3t} \qquad f_2(t) = e^{-4t}\cosh 2t$$

the Laplace transforms are given by

```
syms t a w
f1 = t^3*exp(-3*t);
f2 = cosh(2*t)*exp(-4*t);
F1 = laplace( f1 )
F2 = laplace( f2 )
```

to obtain the results

```
F1 =
     6/(s+3)^4
F2 =
     (s+4)/(s+6)/(s+2)
```

For the inverse Laplace transforms we have that

```
g1 = ilaplace(F1), g2 = ilaplace(F2)
```

g1 =

   t^3*exp(-3*t)

g2 =

   cosh(2*t)*exp(-4*t)

That is, we obtain the original functions.

A property for Laplace transforms says that the derivative of a function has the Laplace transform

$$\mathcal{L}\left\{\frac{df(t)}{dt}\right\} = sF(s) - f(0) \tag{9.4}$$

If we have a function with vanishing initial conditions, then we can use the instruction laplace to find the transform of a derivative. For example, for the function

$$f(t) = t\,\sin(t)$$

its Laplace transform is

F = laplace(f)

F =

2*s/(s^2+1)^2

then, the Laplace transform of its derivative is

laplace(diff(f))

ans =

   2*s^2/(s^2+1)^2

We can clearly see that the Laplace transform of f(t) is multiplied by s when we first take the derivative of f(t) and then the Laplace transform.

Example 9.4 Fourier transforms

In a similar way to the evaluation of Laplace transforms, we can obtain the Fourier transform of a function as well as its inverse Fourier transform. The Fourier transform of a function y(t) is defined by

$$\mathcal{F}\{y(t)\} = Y(j\omega) = \int_{-\infty}^{\infty} y(t)e^{-j\omega t}\,dt \tag{9.5}$$

and the inverse Fourier transform is

$$\mathcal{F}^{-1}\{F(j\omega)\} = f(t) = \frac{1}{2\pi}\int_{-\infty}^{\infty} F(j\omega)e^{j\omega t}\,d\omega \tag{9.6}$$

The MATLAB instructions to get the Fourier and the inverse Fourier transforms are **fourier** and **ifourier**, respectively. For example, for the function

$$f(t) = e^{-|t|}$$

its Fourier transform is then obtained with

```
syms x
f = exp(-3*abs(x));
fourier(f)
```

```
ans =
     6/(w^2+9)
```

As another example, for the function $F(w)=\pi(w\text{-}3)$, the inverse Fourier transform is

```
F = pi*(w-3);
f = ifourier(F)
```

```
f =
    -pi*(i*dirac(1,x)+3*dirac(x))
```

where the dirac function is the impulse function at x=+1 for **dirac(1,x)** and at x=0 for **dirac(x)**.

Example 9.5 z-transform

The z-transform is defined for discrete signal sequences $x(n)$ and it is given by

$$\mathcal{Z}\{x(n)\} = X(z) = \sum_{n=-\infty}^{\infty} x(n)z^{-n} \tag{9.7}$$

The inverse z-transform is given

$$x(n) = \mathcal{Z}^{-1}\{X(z)\} = \frac{1}{2\pi j}\oint_C X(z)z^{n-1}\,dz \tag{9.8}$$

They can be evaluated with the instructions **ztrans** and **iztrans**, respectively. As an example let us consider the sequence $x(n) = e^{-kn}$. Its z-transform is

```
syms k n
x = exp(-k*n);
X = ztrans(x)
```

```
X =
    z/exp(-k)/(z/exp(-k)-1)
```

the result can be rewritten as

$$X(z) = \frac{z}{z - e^{-k}}$$

For an example in the evaluation of an inverse z-transform, let us consider the z-transform given by

$$X(z) = e^{\frac{1}{z}}$$

The inverse z-transform is obtained with

```
syms z
X = exp(1/z);
iztrans(X)
```

```
ans =
    1/n!
```

That is, the original sequence is

$$x(n) = \frac{1}{n!}$$

9.2   Applications in Digital Signal Processing

The area of digital signal processing has also benefited from the use of the Signal Processing toolbox. We show some examples for discrete signals. We start with a convolution, then continue with the discrete Fourier transform and how we can perform convolution with it. Finally we design some digital filters with this toolbox.

Example 9.6 Linear convolution

The linear convolution of two finite-duration (also called finite length sequences) discrete signals $x_1(n)$, and $x_2(n)$ is given by

$$y(n) = \sum_{k=-\infty}^{\infty} x_1(n)x_2(k - n) \tag{9.9}$$

MATLAB performs a convolution with the instruction conv(x1, x2). The values of the elements of the sequences must be in a polynomial format. Thus, for the sequences

$$x1 = [1\ 2\ 3\ 4\ 5\ 4\ 3\ 2\ 1] \quad \text{and} \quad x2 = [1\ 2\ \text{-}1]$$

we have that the convolution is a new sequence $y(n)$ of length $n_1 + n_2 - 1$ and given by

x1 = [1 2 3 4 5 4 3 2 1]; x2 = [1 2 -1];
y = conv(x1, x2)

y =
     1 4 6 8 10 10 6 4 2 0 -1

we see that the result has a length 11 because x_1 and x_2 have lengths 9 and 3, respectively. We can plot the convolution sequence with

stem (y)
xlabel (' n')
ylabel('y(n)')
axis([0 12 -3 12])

The convolution sequence is plotted in Figure 9.4.

Example 9.7 Discrete Fourier transform

The discrete Fourier transform (DFT) of a sequence of length N is given by

$$X(k) = \sum_{n=0}^{N-1} x(n)e^{-j\frac{2\pi kn}{N}} \qquad k = 0, 1, ..., N-1 \qquad (9.10)$$

The inverse discrete Fourier transform (IDFT) is given by

$$x(n) = \frac{1}{N}\sum_{k=0}^{N-1} X(k)e^{j\frac{2\pi kn}{N}} \qquad n = 0, 1, ..., N-1 \qquad (9.11)$$

Actually, MATLAB computes the DFT using the algorithm for the fast Fourier transform (FFT) [1]. This algorithm is an efficient method to compute the DFT of a sequence. It optimizes the number of multiplies and additions used in the computation of the DFT. The instructions fft and ifft are used to perform the DFT and the IDFT in MATLAB. For example, for the finite length sequence

$$x_1(n) = \begin{cases} 1 & 0 \le n < 5 \\ 0 & 5 \le n \le 9 \end{cases}$$

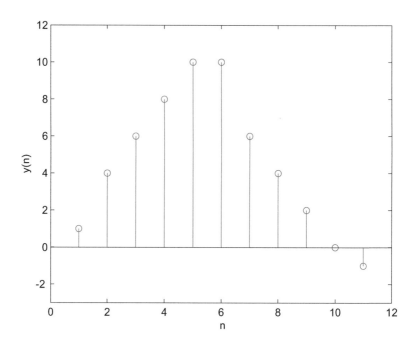

Figure 9.7: Sequence x_3 resulting from the convolution of x_1 and x_2.

we have written the following script to calculate and plot $x_1(n)$ and the DFT:

```
% This is file Example9_7a.m
% Computes the DFT of a sequence and then it plots
% the sequence and the DFT.
clear
clc
close all
x1 = [ 1 1 1 1 1 0 0 0 0 0 ];% Defines the sequence.
subplot ( 2, 1, 1 ) % Defines the subplot for x1.
stem( x1 ) % Plots the sequence x1.
axis([0 10 -1 2])
xlabel('n')
ylabel('Sequence x_1(n)')
X1 = fft ( x1 );% Evaluates the DFT.
k = [ 0 : 1 : 9 ]; % Defines the vector from 0 to 9.
subplot ( 2, 1, 2 ) % Defines the subplot for the DFT.
stem ( k, X1 ) % Plots the DFT of x1.
```

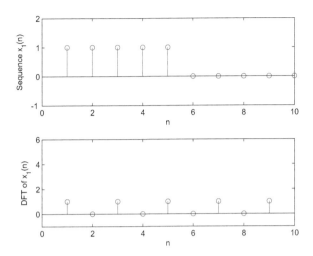

Figure 9.8: Sequence $x(n)$ and its DFT.

```
axis([0 10 -1 6])
xlabel('n')
ylabel('DFT of x_1(n)')
stem ( k, X1 ) % Plots the DFT of x1.
```

The sequence $x_1(n)$ and its DFT $X(k)$ are shown in Figure 9.8.

The circular convolution of two sequences is obtained by multiplying the DFT of the sequences. Thus for the sequences $x_1(n)$ and $x_2(n)$ of Example 9.6 we have

```
x1 = [1 2 3 4 5 4 3 2 1];
x2 = [1 2 -1];
F1 = fft(x1)

F1 =
Columns 1 through 3
25.0000 -7.7909 - 2.8356i 0.2169 + 0.1820i

Columns 4 through 6
-0.5000 - 0.8660i 0.0740 + 0.4195i 0.0740 - 0.4195i

Columns 7 through 9
-0.5000 + 0.8660i 0.2169 - 0.1820i -7.7909 + 2.8356i
```

F2 = fft(x2)

F2 =

2.0000 0.5000 - 2.5981i 0.5000 + 2.5981i

We see that the two DFTs have different lengths and in order to multiply them they must have the same length, thus we pad F2 with zeros in such a way that it has the same length as that of F1. F2 is now given by

F2 = 2.0000 0.5000 - 2.5981i 0.5000 + 2.5981i 0 0 0 0 0 0 0

which is of length 9. Now we just multiply them as

F3 = F1.*F2

ans =

50.0000 -11.2627 +18.8236i -0.3644 + 0.6545i 0 0 0 0 0 0

We finally take the IDFT to obtain the convolution as

f3 = ifft(F3)

f3 =
Columns 1 through 3
4.2637 + 2.1642i 3.1739 + 0.7706i 3.2917 - 0.9514i

Columns 4 through 6
4.4532 - 2.1308i 6.0319 - 2.3117i 7.3691 - 1.5077i

Columns 7 through 9
7.9498 - 0.0334i 7.4609 + 1.5411i 6.0059 + 2.4591i

Note that the sequence f3 has length 9, as opposed to the result in Example 9.6. This is so because the circular convolution is different from the linear convolution. However, we can perform linear convolution with the DFT just by padding with zeros the two sequences adding as many zeros as the final length of the linear convolution. Thus, in the case of the same sequences we pad x2 with zeros so now they become

$$x1 = [1\ 2\ 3\ 4\ 5\ 4\ 3\ 2\ 1\ 0\ 0];$$
$$x2 = [1\ 2\ -1\ 0\ 0\ 0\ 0\ 0\ 0\ 0\ 0];$$

Now we perform the following computations to obtain the linear convolution of $x_1(n)$ and $x_2(n)$,

X1 = fft(x1); X2 = fft(x2);
X1.*X2;
X3 = X1.*X2

X3 =
Columns 1 through 3
50.0000 -19.9268 -19.7614i 0.1898 + 0.7092i

Columns 4 through 6
0.2700 - 4.0678i -0.1922 + 0.8645i 0.1592 - 1.2008i

Columns 7 through 9
0.1592 + 1.2008i -0.1922 - 0.8645i 0.2700 + 4.0678i

Columns 10 through 11
0.1898 - 0.7092i -19.9268 +19.7614i

x3 = ifft(X3)

x3 =

Columns 1 through 6
1.0000 4.0000 6.0000 8.0000 10.0000 10.0000

Columns 7 through 11
6.0000 4.0000 2.0000 -0.0000 -1.0000

We can see that the linear convolution obtained using the DFT is the same as the one obtained using the instruction conv.

9.3 Applications in Control

MATLAB finds many applications in Control theory. Thus, a specially control-dedicated toolbox has been produced and is available as a separate product. In this section we present some examples in Control theory.

Example 9.8 Stability in a feedback system

MATLAB can perform a great deal of functions from the Control toolbox. For example, let us consider the feedback system shown in Figure 9.9. The plant is defined by the function

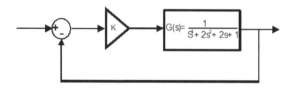

Figure 9.9: Block diagram of a plant with feedback.

$$G(s) = \frac{1}{s^3 + 2s^2 + 2s + 1}$$

This plant can be defined in MATLAB as

nump = [1];
denp = [1 2 2 1];
plant = tf(nump, denp) % This instruction writes the transfer function.

and MATLAB delivers

Transfer function:
         1
_____

s^3 + 2 s^2 + 2 s + 1

If the system gain is K=1, the negative feedback system is defined as

system = feedback(plant, [1])

to obtain the complete system transfer function as

system = feedback(plant, [1])

Transfer function:
         1
_____

s^3 + 2 s^2 + 2 s + 2

(Note the independent coefficient in the denominator is now 2.) To check for system stability we obtain the poles with

poles = pole(system)

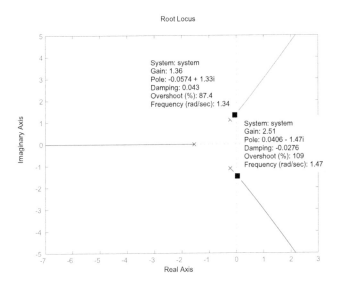

Figure 9.10: Root locus of a feedback system.

to get

```
poles =
-1.5437
-0.2282 + 1.1151i
-0.2282 - 1.1151i
```

We see that the system has three poles because we have a third-order denominator and that the three poles are in the left-hand plane and thus, the system is stable. We can find out more on this system. For example, we can obtain its root locus. This is a plot of root loci versus the gain K. A root locus can be found with the instruction rlocus as in

rlocus(system)

The root locus is shown in Figure 9.10. We show two data tips and we see the conditions for stability and instability. We see that a gain of 2.36 makes the system unstable.

Example 9.9 Control of the Hubble telescope

A model for the Hubble telescope which is in orbit around the Earth, together with a positioning system is shown in Figure 9.11.

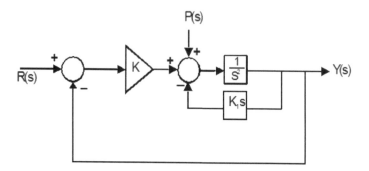

Figure 9.11: Model for the Hubble telescope.

The goal is to find values for $K1$ and $K2$ such that the overshoot when a step input is applied is less than or equal to 5%, that the error when we apply a ramp $r(t)$ is minimized, and that the effect of a perturbation step is reduced. The system has two inputs, $R(s)$ and $P(s)$, and then we have

$$Y(s) = \frac{KG(s)}{1 + KG(s)}R(s) + \frac{G(s)}{1 + KG(s)}P(s) \tag{9.12}$$

The error is given by

$$E(s) = \frac{1}{1 + KG(s)}R(s) - \frac{G(s)}{1 + KG(s)}P(s) \tag{9.13}$$

First we try to satisfy the 5% overshoot requirement. Thus, we only take into account the input-output system with a step input with value A as:

$$\frac{Y(s)}{R(s)} = \frac{KG(s)}{1 + KG(s)} = \frac{K}{s(s + K_1) + K} = \frac{K}{s^2 + K_1 s + K} \tag{9.14}$$

From Figure 9.5, we see that for a 5% overshoot we need $\xi = 0.7$. For a ramp $r(t) = Bt$, the steady-state error is given by

$$\epsilon_{ss}^r = \lim_{s \to 0} \frac{B}{KG(s)} = \frac{B}{K/K_1}$$

And for a unit step perturbation

$$\epsilon_{ss}^r = \lim_{s \to 0} \frac{-1}{(s + K) + K} = \frac{-1}{K}$$

In order to decrease the error due to a perturbation we need to increase K. In addition, to decrease the steady-state error when the input is a ramp we need to increase K/K_1. Further we need $\xi = 0.7$. The characteristic equation is

$$s^2 + 2\xi\omega_n s + \omega_n^2 = s^2 + k_1 s + K = s^2 + 2(0.7)\omega_n s + K$$

Thus, $\omega_n = \sqrt{K}$ and $K_1 = 0.7\omega_n$, and then $K_1 = 1.4\sqrt{K}$. Therefore,

$$\frac{K}{K_1} = \frac{K}{1.4\sqrt{K}} = \frac{\sqrt{K}}{1.4} = \frac{\sqrt{K}}{2\xi}$$

If $K=36$, $K_1 = 1.4 \times 6$ and $K/K_1=36/8.4$, and if $K=100$, $K_1=1.4$, the step, the ramp, and the perturbation responses are shown in Figure 9.12 and can be obtained with the following script

```
% This is file Example9_9.m
% We analyze the behavior of the Hubble telescope
clear
clc
close all
% We first evaluate the plant
K = 100;
K1 = 14;
nump = [ 1 ];
denp = [ 1 K1 0 ];
% Transfer function for the plant
plant = tf( nump, denp )
% Transfer function from r(t)
system = feedback ( K*plant, [ 1 ] )
% Perturbation
% Transfer function from the perturbation
perturbation = feedback ( plant, [ K ] )
step ( system )
hold on
step ( perturbation )
hold off
```

We can see from Figure 9.12 that the specifications are satisfied for the design.

9.4 Applications in Chemical Engineering

Example 9.10 Mass transfer

We wish to find the water vapor flux that evaporates from a vessel with water at 25° at a pressure of 1 atm. The distance from the water surface to the top of the vessel is 0.4 m. The water vapor flux is given by Bird's equation [1,2]

$$N_z = -cD\frac{\partial x}{\partial z} - x\left(\frac{cD}{1-x_0}\right)\frac{\partial x}{\partial z}\bigg|_{z=0} \tag{9.15}$$

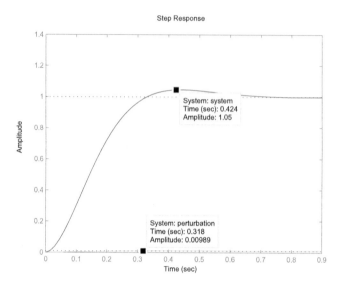

Figure 9.12: Response for the Hubble telescope.

where N_z is the water vapor flux at a point z (the z axis is the vertical one), C is the total concentration in the vapor phase, D is the water diffusion coefficient in the air, x is the molar fraction of the water vapor, and x_0 is the value of x at $z = 0$. The non-steady concentration, assuming that there is no air flow in the vessel opening, can be obtained with

$$\frac{x}{x_0} = C = \frac{1 - erf(Z - \varphi)}{1 + erf\varphi} \tag{9.16}$$

where

$$Z = \frac{z}{\sqrt{4Dt}} \tag{9.17}$$

Here, t is time and φ can be found from the solution of the non-linear equation for N_z,

$$x_0 = \frac{1}{1 + [\sqrt{\pi}(1 + erf\varphi)\varphi e^{\varphi^2}]^{-1}} \tag{9.18}$$

For the system we have the following constants:

$$D = 2.2 \times 10^{-5} \ \mathrm{m}^2/\mathrm{s}$$

$$X_0 = \frac{P^{sat}(@25°C)}{Pi} = 0.0312$$

$$c = \frac{P_t}{RT} = 0.049 \ \text{Kmol/m}^3$$

To solve this problem, first we find φ from Equation (9.18). We then rewrite that equation as

$$\frac{1}{1 + [\sqrt{\pi}(1 + \text{erf } \varphi)\varphi \ e^{\varphi 2}]^{-1}} - x_0 = 0$$

We can now create an m-function for this function as

```
function phi_a = phi_b( x , x0 ) % Non linear equation to solve for phi.
global x0
phi_a = 1/( 1 + (sqrt( pi )*( 1 + erf( x ))*x*exp( x^2 ))^(-1))-x0;
```

The script to calculate the water vapor flux is:

```
% File Example9_10.m
% This file calculates the water vapor flux
% through the reservoir top.
close all
clc
clear
global x0
syms z
x0 = 0.0312;
z0 = 0;
c = 0.0409;
t = [ eps: 20 :3600 ];
z0 = 0.1; % Distance from surface of water to top of vessel.
D = 2.2e-5; % Water concentration.
x0 = 0.0312; % Molar fraction.
% Solution of the equation
phi = fzero( 'phi_b', 1 );
x = x0*(1-erf(z./sqrt(4*D*t)))/(1 + erf( phi));
a = diff( x, z);
zi= [0.1 0.2 0.3 ];
cc = subs( a, z, 0 );
for i = 1:3
    b = subs( a, z, zi(i) );
    xx = subs ( x, z, zi(i) );
    N( i, : ) = -c*D*b-c*D*xx.*cc/(1-x0);
end
plot( t/60, N*3600*18*1000 )
legend( 'z = 0.1' , 'z = 0.2' ,'z = 0.3' )
```

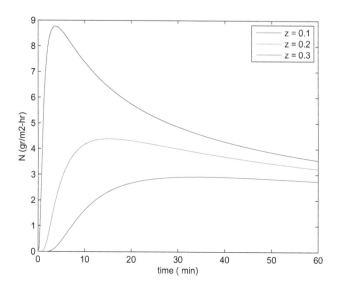

Figure 9.13: Flux versus time.

```
xlabel( 'time ( min)' )
ylabel( 'N (gr/m2-hr)' )
figure
for i= 1:3;
    dd(i,:) = subs( x, z, zi(i) );
end
plot ( t/60 , dd(1,:), t/60 , dd(2,:),'+',t/60 , dd(3,:),'*')
legend( 'z = 0.1' , 'z = 0.3' ,'z = 0.3' )
xlabel( 'time ( min)' )
ylabel( 'Molar fraction for water' )
```

Figures 9.13 and 9.14 show the water vapor flux and the molar fraction.

Example 9.11 Chemical reaction system

Let us consider the chemical reaction

$$A \xrightarrow{k_1,k_2} B \xrightarrow{k_3,k_4} C \qquad (9.19)$$

Where the k_i are the kinetic rate constants. k_1 (k_3) is the kinetic rate constant to change $A \rightarrow B$ $(B \rightarrow C)$ and k_2 (k_4) is the kinetic constant for $B \rightarrow A$ $(C \rightarrow B)$. Let us assume the values for the kinetic constants are $k_1=1$ min^{-1}, $k_2=0$ min^{-1}, $k_3=2$ min^{-1}, $k_4=3$ min^{-1}. The initial concentrations at time $t=0$ are:

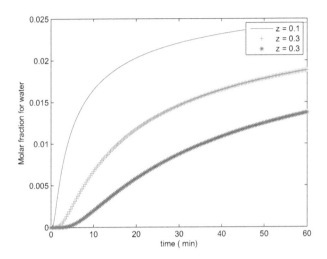

Figure 9.14: Molar fraction.

$$C_A(0) = 1$$

$$C_B(0) = 1$$

$$C_C(0) = 1$$

The rate for each reaction is given by the system of differential equations:

$$\frac{dC_A}{dt} = -k_1 C_A + k_2 C_B$$

$$\frac{dC_B}{dt} = k_1 C_A - k_2 C_B - k_3 C_B + k_4 C_C \tag{9.20}$$

$$\frac{dC_C}{dt} = k_3 C_B - k_4 C_C$$

In matrix form, these equations can be written as

$$\dot{C}(t) = kC(t) \tag{9.21}$$

$$C(t) = \begin{bmatrix} C_A(t) \\ C_B(t) \\ C_C(t) \end{bmatrix} \tag{9.22}$$

$$\dot{C}(t) = \begin{bmatrix} \frac{dC_A(t)}{dt} \\[2mm] \frac{dC_B(t)}{dt} \\[2mm] \frac{dC_C(t)}{dt} \end{bmatrix} \tag{9.23}$$

$$K = \begin{bmatrix} -k_1 & k_2 & 0 \\ k_1 & -k_2 - k_3 & k_4 \\ 0 & k_3 & -k_4 \end{bmatrix} = \begin{bmatrix} -1 & 0 & 0 \\ 1 & -2 & 3 \\ 0 & 2 & -3 \end{bmatrix} \tag{9.24}$$

To solve this system of differential equations we define them in a function as

```
function Cpunto = dc( t, C)
global k
Cpunto( 1 ) = -k( 1 )*C(1) + k(2)*C(2) ;
Cpunto( 2 ) = +k(1)*C(1)-k(2)*C(2) - k(3)*C(2) + k(4)*C(3);
Cpunto( 3 ) = k( 3 )*C( 2 ) - k( 4 )*C( 3 );
Cpunto=Cpunto';
```

With the following script we solve the equations and plot the concentration profiles:

```
% This is file Example9_11.m
% It computes the component concentration profiles
% in a chemical reaction.
global k
% Input data:
fprintf( 'Enter the kinetic constants')
k( 1 ) = input (' A->B , k1 = ');
k( 2 ) = input (' B->A , k2 = ');
k( 3 ) = input (' B->C , k3 = ');
k( 4 ) = input (' C->B , k4 = ');
fprintf( 'Enter the initial concentrations' )
C0( 1 ) = input( ' Initial concentration for A = ' );
C0( 2 ) = input( ' Initial concentration for B = ' );
C0( 3 ) = input( ' Initial concentration for C = ' );
tmax = input( ' Enter the maximum time tmax = ');
tmax = 5;
k = [1, 0, 2, 3];
C0 = [1 0 0];
t = [ 0: 0.01: tmax ];
[ t, C ]= ode45( 'dc' , [ 0, tmax], C0);
plot(t, C)
```

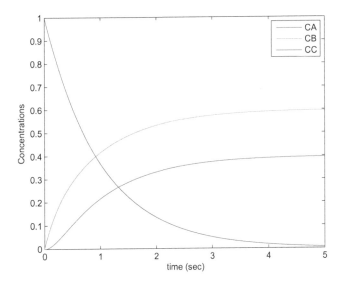

Figure 9.15: Concentration profiles for the components.

```
xlabel( ' time (sec)' )
ylabel( 'Concentrations' )
legend( 'CA', 'CB' , 'CC' )
```

Figure 9.15 shows the concentration profiles. Note that at any time the principle of conservation of matter holds. That is,

$$C_A(0) + C_B(0) + C_C(0) = C_A(t) + C_B(t) + C_C(t) \qquad (9.25)$$

9.5 Applications in Industrial Engineering

Industrial engineering makes extensive use of mathematics. For example, optimization of industrial processes, statistical analysis, cost analysis, linear programming, differential equations, among many other techniques. In this section we present examples showing how the MATLAB optimization package can be used to optimize a goal function and to minimize cost in a process. This section requires the use of the Optimization toolbox.

Example 9.12 Cost optimization

In a water treatment plant, one of the most important processes is the removal of turbidity by adding chemical substances that group together the solids suspended, thus forming larger particles that can be removed by sedimentation. Then, those particles are removed by filtration. In general, the greater the amount of chemicals, the greater number of particles is formed. This means that the filter washing frequency decreases and in this case the washing costs decrease too. We wish to know what is the optimal amount of chemicals that have to be used to optimize the total cost of the process. The turbidity and the chemicals are experimentally related according to Table 9.1.

Table 9.1: Experimental data

Chemicals	Turbidity
1	31
2	28
3	19
4.5	14
7.5	9
13	8
20	5
27	8

Using polynomial interpolation with instruction **polyfit** we obtain a second-degree polynomial that interpolates the experimental data. Thus, for $n=2$ we can use

Q = [1 2 3 4.5 7.5 13 20 27];
T = [31 28 19 14 9 8 5 8];
P = polyfit(Q, T, 2)

MATLAB produces the coefficient polynomials

P =

0.0856 -3.0894 30.5223

corresponding to the polynomial

$$T = 0.0856Q^2 - 3.0894Q + 30.5223$$

where Q is the amount of chemicals and T is the turbidity. The rate of water needed for the filter cleaning is B m^3/1000 m^3 and is given by the equation

$$B = -0.5 + 1.7T$$

Sustituting here the turbidity equation and multiplying by the cost of water which is $0.05/m^3$ we obtain the water washing filter cost C_B in $/1000 m^3$ as

$$C_B = 0.0073Q^2 - 0.2626Q + 2.5694$$

If the cost of the chemicals is

$$C_C = 0.75Q$$

the total cost is

$$C_T = C_B + C_C$$

The following script plots the equations for water cost C_B, C_C, and C_T. The plot in Figure 9.16 shows that the cost is minimized for $Q = 11.8$.

```
% This is file Example9_12.m
% It plots the costs to find out what is the
% optimal amount of chemicals.
clear
close all
clc
Q = [ 1 2 3 4.5 7.5 13 20 27];
T = [ 35 28 19 14 9 5 2 1 ];
a = polyfit( Q, T , 3 )
q = [1 : 0.1 :27];
aa=polyval(a, q);
Cb= -0.5 + 1.7*aa;
Cc = 0.75*q;
CT = Cb + Cc;
plot( q, CT, q, Cb, q, Cc)
grid on
legend( 'CT', 'CB', 'Cc' )
xlabel( 'Chemicals' )
```

Example 9.13 Goal programming

In goal programming we do not look for a maximum or minimum of a function (extremal points) but rather we try to minimize the deviations between the desired goals and the real-life results according to a set of priorities [3,4]. That is, we try to minimize the objective function Z given by

$$Z = \sum w_i(d_i^+ + d_i^-) \tag{9.26}$$

under the restrictions

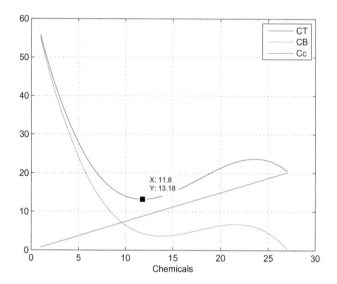

Figure 9.16: Plot for total cost, chemicals cost, and filter washing water cost.

Table 9.2: Hours required per team

Rooler skates	Assembly	Finishing	Net profit
Type A	1	1	15
Type B	3	1	25
Man-Hour/day/dept.	60	40	—

$$\sum a_i x_i + d_i^- - d_i^+ + b_i \quad \forall \ i \qquad (9.27)$$

$$x_j, d_i^-, d_i^+ \geq 0 \quad \forall \ j$$

Now we assume that the plant wants to produce two types of roller skates, namely, type A and type B roller skates. The net profit in the type A skates is \$25.00 whereas that the net profit in the type B ones is \$15.00. The unit assembly and finishing times are shown in Table 9.2. If we wish to optimize this procedure we note that we must maximize

$$Z = 15x_1 + 25x_2 \qquad (9.28)$$

with the restrictions

$$x_1 + 3x_2 \leq 60 \qquad \text{(Assembly)}$$
$$x_1 + x_2 \leq 40 \qquad \text{(Finishing)}$$

This problem can be solved in MATLAB by minimizing $-Z$. First we define Z in a function as

```
function f = objfun(x)
f = -15*x(1) - 25*x(2);
```

The script to optimize the function is:

```
clear
clc
close all
x0 = [ 1 ; 1 ]; % Initial value for x.
options = optimset('LargeScale','off');
lb = [ 0 ; 0 ]; % Lower bounds.
ub = [ 50 ; 50 ]; %Upper bounds.
A = [1 3 ; 1 1 ];
b = [ 60 ; 40 ];
% Optimization with the function fmincon from
% the optimization package.
[x, fval] = fmincon(@objfun, x0, A, b,[ ],[ ], lb, ub)
```

After running the script we get

```
x =

    30.0000
    10.0000

fval =

    -700
```

This result indicates that the net profit is $700.00, and the production must be of 30 two-wheel roller skates and 10 four-wheel ones. On the other hand, if a minimum net profit of $600.00 we have a problem in goal programming. We then add two deviation variables

$$d_1^+ = \text{goal above the net profit.}$$

$$d_1^- = \text{goal below the net profit.}$$

The net profit is

$$15x_1 + 25x_2 + d_1^- - d_1^+ = 600 \leq 40 \qquad \text{(Net profit)}$$

Then we must minimize

$$Z = d_1^- + d_1^+ \qquad (9.29)$$

with the restrictions

$$x_1 + 3x_2 \leq 60 \qquad \text{Hours-assembly}$$

$$x_1 + x_2 \leq 40 \qquad \text{Hours-finishing}$$

$$15x_1 + 25x_2 + d_1^- - d_1^+ = 600 \qquad \text{Net profit}$$

$$x_1, x_2, d_1^-, d_1^+ \geq 0$$

The following function evaluates Z,

```
function f = objfunB(x)
f = x(3)+ x(4);
```

and the following script finds x_1 and x_2,

```
% This is file Example9_13B.m
% Solves the Goal programming problem.
clear
clc
close all
x0 = [ 1 ; 1 ; 1 ; 1 ]; % Valor inicial
options = optimset('LargeScale','off');
lb = [1;0; 1;1];
ub = [50;50;50;50];
A = [1 3 0 0; 1 1 0 0 ; -1 0 0 0; 0 -1 0 0 ; 0 0 -1 0; 0 0 0 -1];
b = [ 60 ; 40 ; 0 ; 0 ; 0; 0 ];
Ae = [ 15 25 -1 1 ; 0 0 0 0; 0 0 0 0; 0 0 0 0; 0 0 0 0; 0 0 0 0];
be = [600; 0; 0; 0; 0; 0];
x = fmincon(@objfunB, x0, A, b, Ae, be, lb, ub)
```

MATLAB gives the optimal solution

```
x =
    15.0000
    15.0000
    1.0000
    1.0000
```

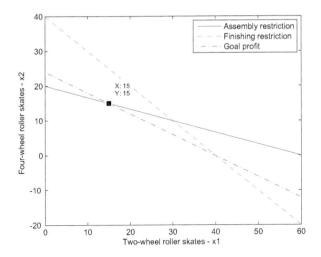

Figure 9.17: **Graphical method to solve the goal programming problem.**

A graphical method to plot the solution can be done with the following script. The results are shown in Figure 9.17 which corresponds to assembly and finishing and the net profit.

```
% This is file Example9_13C.m
% Plots the solution for goal programming.
clear
clc
close all
xx1 = [ 0 : 5 : 60 ];
xx2 = 60/3-xx1/3;
xx3 = 40-xx1;
xx4 = 600/25-xx1*15/25;
plot( xx1, xx2, xx1, xx3, '-',xx1, xx4,'-.' )
xlabel( 'Two-wheel roller skates - x1' )
ylabel( 'Four-wheel roller skates - x2' )
legend( 'Assembly restriction','Finishing restriction','Goal profit')
```

We readily see that the solution found with the graphical method is the same as the optimum solution. Note that $15*x(1)+25*x(2) = 600$ thus satisfying the profit goal.

9.6   Applications in Food Engineering

In this section we present two examples which are easily soved with MATLAB.
The first one is a data-fitting problem to appreciate the behavior of specific
heat from experimental data. The second one evaluates the behavior of a
series solution when the number of terms in the series solution is different.

Example 9.14 3-D plotting of experimental data

During the past decade, heat and mass transfer during the frying of meat
products have been widely studied. In a recent paper [4], the deep frying of
pork meat has been studied and some experimental results have been presented. In this paper, the authors have presented data for the specific heat
in a table. In this example we plot in a 3D plot the same data so the reader
can have a better feeling of the experimental data. The data from the paper
is tabulated and is shown in Table 9.3. Here we see that there are four frying
temperatures and four frying times. The script shown below realizes the 3D
mesh plot. Figure 9.18 shows the 3D mesh plot and there we can see the
behavior of the process.

```
% This is file Example9_14.m
% It plots the specific heat for pork meat in the frying process.
clear
close all
t = 0: 30: 120;
% Cp data at 90 degrees
Cp(1, :) = [4.54, 3.56, 3.54, 3.36, 3.13];
% Cp data at 100 degrees
Cp(2, :) = [3.99, 3.78, 3.35, 3.31, 2.78];
% Cp data at 110 degrees
Cp(3, :) = [3.72, 3.58, 3.17, 2.85, 2.77];
% Cp data at 120 degrees
Cp(4, :) = [4.62, 4.43, 3.76, 3.46, 3.41];
T=[90, 100, 110, 120]
[x, y] = meshgrid(t, T);
mesh(x, y, Cp)
axis([0 120 90 120 2 5])
xlabel('time')
ylabel('Temperature ^\circ C')
zlabel('Specific heat C_p')
```

Example 9.15 Fick's law

Every diffusion process is governed by Fick's law. If the diffusion is in a
steady-state, Fick's first law establishes that the concentration profile $c(x)$

Table 9.3: Data for specific heat

$T(°C)$ (kJ/kgC)	$t(min)$	C_P
90	0	4.54
	30	3.56
	60	3.54
	90	3.36
	120	3.13
100	0	3.99
	30	3.78
	60	3.35
	90	3.31
	120	2.78
110	0	3.72
	30	3.58
	60	3.17
	90	2.85
	120	2.77
120	0	4.62
	30	4.43
	60	3.76
	90	3.46
	120	3.41

and the diffusion flux J, which is the mass transported by unit area by unit time are related by

$$J = -D\frac{dC}{dx} \qquad \text{Fick's first law} \qquad (9.30)$$

where D is the diffusion coefficient. In this equation we assume that the diffusion takes place in the x direction. If the diffusion is not in the steady-state, then the diffusion process is governed by Fick's second law,

$$\frac{\partial C}{\partial t} = -D\frac{\partial^2 C}{\partial x^2} \qquad (9.31)$$

where $C = C(x,t)$. A solution for this equation is given by:

$$\frac{C_x - C_0}{C_s - C_0} = \frac{6}{\pi^2}\sum_{n=1}^{\infty}\frac{1}{n^2}exp\left(-n^2\frac{D\pi^2}{r^2}\right) \qquad (9.32)$$

where C_s is the surface concentration of the material diffusing, C_0 is the initial concentration, C_x is the concentration at a distance x from the surface,

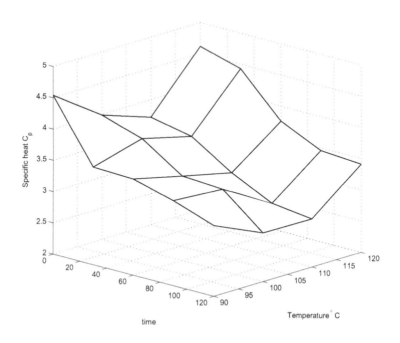

Figure 9.18: 3D plot for specific heat versus temperature and time.

D is the diffusion coefficient, and r is the diffusion depth. In particular, we wish to know the humidity content at beet after it has been subjected to a drying process for a time t. The beet has been cut and shaped in spheres with a 2.5-cm diameter. The drying furnace has a temperature of 60° the equilibrium humidity is 0.066 and the initial humidity is 0.25 $\mathrm{kg_{water}/kg_{ss}}$. The beet diffusion coefficient is 7×10^{-10} m^2/s. The following m-file plots the humidity content during the first 60 seconds. We plot in Figure 9.19 the results when we plot the solution in the series of Equation 9.32, plotting from a single term in the series up to 13 terms. We see that this series converges very fast.

```
% This is file Example9_15.m
% Plots the drying process of beet
% according to Fick's law.
clear
clc
close all
y = zeros(11, 3601);
Xi = 0.25;
Xe = 0.066;
```

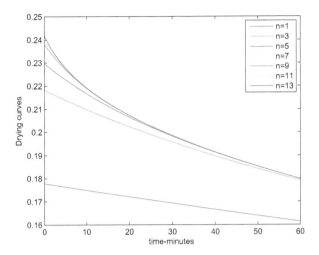

Figure 9.19: Drying curves for different numbers of terms in the series.

```
x1 = (6/pi^2);
t = [0: 1: 3600];
D = 7e-10;
r = 1.25e-2;
x = 0;
for n = 1: 13;
    x = x+x1*(1/n^2)* exp(-n^2*D*pi^2*t/r^2);
    X = x*(Xi-Xe)+ Xe;
    y(n,:) = X;
end
plot ( t/60, y(1:2:13,:))
hold on
legend ( 'n = 1','n = 3','n = 5','n = 7','n = 9','n = 11', 'n = 13' )
ylabel( 'Drying curves' )
xlabel( 'time-minutes'
```

9.7 Applications in Civil Engineering

In this section we present two examples where MATLAB can be used. The first example is a beam deflection problem with a uniformly distributed load. The second problem is support reaction problem. Both examples are easily solved with MATLAB.

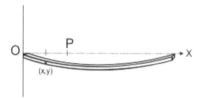

Figure 9.20: Deflected beam.

Example 9.16 Beam deflection

Let us consider a horizontal beam with length $L = 20$ m supported on both ends. It has a load uniformly distributed $w = 100$ Kg/m. We wish to find the equation that describes the beam deflection.

Let us suppose that the left end is the origin of the coordinate system as shown in Figure 9.20. In the origin end, the beam has a forward upward vertical force given by $w \times L = 100 \times 20$ Kg. For any point P with coordinates (x, y) along the beam we have a load in the middle point of the segment OP given by wx. The differential equation is

$$E \cdot I \cdot \frac{d^2 y}{dx^2} = wLx - wx\left(\frac{1}{2}x\right) = wLx - \frac{1}{2}wx^2 \tag{9.33}$$

where E is the elasticity modulus and I is the moment of inertia of a tranversal section. This differential equation can be solved in MATLAB by simply integrating twice with respect to x from $x = 0$ to $x = 20$. To integrate we can use the instruction int but we first rewrite the differential equation as

$$\frac{d^2 y}{dx^2} = \frac{wLx}{EI} - \frac{wx}{EI}\left(\frac{1}{2}x\right) = w\left(Lx - \frac{x^2}{2}\right)\frac{1}{EI} \tag{9.34}$$

In MATLAB we solve this as

```
d2y = w*(L*x-x^2/2)/(E*I)
```

The first integral is obtained with

```
dy = int(d2y)
```

and the second one with

```
dy = int(dy)
```

then we find the two constants of integration. If $E = 1000$, $I = 100$, $w = 200$ and $L = 10$, the complete script is

```
% This is file Example9_16.m
% This file integrates the differential equation for the moments.
% E is the elasticity model. I is the moment of inertia.
clc
clear
close all
syms x E I w L
% Differential equation.
d2y = w*(L*x-x^2/2)/(E*I);  % First integral,
dy = int( d2y )
y = int( dy )
% Evaluation of the constants of integration:
% y = 0 at x = 0; y = 0 at x = 20.
C2 = 0;
C1 = -w*L^3/(E*I)/3;
y = y + C1*x;
fprintf ( ' The solution is y =' )
pretty( y )
x1 = [ 0:1:20 ];
% Substitution of normalized values for E and I.
% Substitution of w =100, L=10.
y1 = subs( y, [ E, I, w, L], [ 1000, 100, 100, 10 ] );
y2 = subs( y1, 'x', x1 );
% Maximum deflexion at the center of the beam.
ymx = 5*w*L^4/( 24*E*I );
ymax = subs( ymx, [ E, I, w, L], [ 1000, 100, 100, 10 ])
plot ( x1, y2 )
```

After running the script we get

The solution is

y =

$$
\frac{w\,(1/6\,L\,x^3 - 1/24\,x^4)}{E\,I} - \frac{1/3\,w\,L\,x^3}{E\,I}
$$

ymax =
           2.0833

We also get the plot for the beam deflection shown in Figure 9.10. There we see the maximum beam deflection.

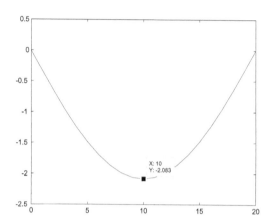

Figure 9.21: Beam deflection.

Example 9.17 Support reactions

Let us consider the support shown in Figure 9.22. To find out the reactions we use force equilibrium. The applied load is q0. At point A, the only movement is a change in the angle θ and there are reactions at point A. At point C, there might be a movement that would change the coordinates (x, y) for point C, besides the reaction at point C as is shown in Figure 9.23. The forces and reactions present at the support are shown in Figure 9.24. The equilibrium equation at point A is

$$\sum M = 0 = -Fh - (2q_0L)(L_2) + R_e h \cos\theta_1 + R_{Ax} \qquad (9.35)$$

Furthermore, we have that $h = L_2 \tan\theta_2$. At point C, the components (x, y) for R_C are

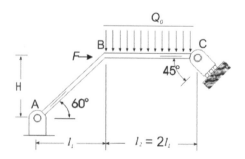

Figure 9.22: Support.

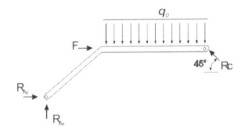

Figure 9.23: Reactions at the support.

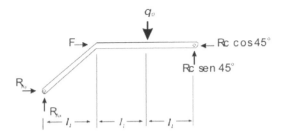

Figure 9.24: Components of the reactions.

$$\sum F_x = -F - R_C \cos\theta_1 + R_{Ax}$$
$$\sum F_y = 0 = R_{Ay} + R_C \sin\theta_1 - q_0 L_2$$
(9.36)

Using the instruction solve we can find the solutions. The following script produces the desired output:

```
% This is file Example9_17.m
% Evaluates the displacement of a structure.
clc
clear
close all
syms L F q0 theta1 theta2 fi R rAx rax RAx RAy rAy L2
fprintf( 'Reaction at point C ' )
r = solve( -F*L*tan(theta2)-( 2*q0*L*L2 )...
    + R*cos(theta1)*L*tan(theta2)+ R*sin(theta1)*3*L, R );
simplify ( r );
expand( ans );
pretty( ans )
r1 = subs( r, theta2, pi/3 );
r3 = subs( r1, theta1, pi/4 );
r2 = subs( r3, L2, 2*L );
simplify( r2 );
```

```
expand( ans );
fprintf('Reaction at point C for theta1 = 60, theta2 = 45 and L2=2*L')
pretty( ans )
RAx = subs( rAx, teta1, pi/4 );
expand( RAx);
simplify( ans);
fprintf('x component for reaction at point A for theta1 = 60 ...
    and theta2 = 45')
pretty( ans)
ray = solve( rAy + r2*sin(teta1) - q0*L2, rAy );
ray1 = subs( ray, teta1, pi/4 );
RAy = subs( ray1, L2, 2*L );
simplify( RAy );
expand( ans );
fprintf('y component for reaction at point A for theta1 = 60 ...
    and theta2 = 45')
pretty( ans )
```

The output is:

Reaction at point C

$$\frac{F\ \sin(\text{theta2})}{\cos(\text{theta1})\ \sin(\text{theta2}) + 3\ \sin(\text{theta1})\ \cos(\text{theta2})}$$

$$+ 2\ \frac{q0\ L2\ \cos(\text{theta2})}{\cos(\text{theta1})\ \sin(\text{theta2}) + 3\ \sin(\text{theta1})\ \cos(\text{theta2})} \text{--}$$

Reaction at point C for theta1 = 60 and theta2 = 45 and L2=2*L

$$\frac{2^{1/2}\ F3^{1/2}}{3^{1/2} + 3} + 4\ \frac{2^{1/2}\ q0\ L}{3^{1/2} + 3}\text{-}$$

x component at point A

$$\frac{-F3^{1/2} - 3F^{1/2} + 2\ 2^{1/2}\ \cos(\text{theta1})\ F3^{1/2} + 42\ \cos(\text{theta1})\ q0\ L}{3^{1/2} + 3}$$

x component of reaction at point A for theta1 = 60 and theta2 = 45

$$\frac{-3\ F + 4\ q0\ L}{\dfrac{1/2}{3} + 3}$$

y component for reaction at point A for theta1 = 60 and theta2 = 45

$$-\frac{\dfrac{1/2}{F3}}{\dfrac{1/2}{3} + 3} + 2\ \frac{\dfrac{1/2}{q0\ L}}{\dfrac{1/2}{3} + 3} + 2\ \frac{q0\ L3}{\dfrac{1/2}{3} + 3}$$

We can see that the outputs produced are expressions for the components for R_{Ax} and R_{Ay}. From here we can conclude that point C will move to the sides depending upon the magnitude of force F.

9.8   Applications in Mechanical Engineering

This section presents two examples that are solved by changing the differential equation to a system of linear differential equations. These sets of equations are then used with the instruction ode23 to give numerical results that can be plotted.

Example 9.18 Damped system under a harmonic movement at the base

Sometimes, the base of a mass-spring-damper system is under a harmonic movement. Figure 9.25 shows the system and Figure 9.26 shows the effect of the harmonic movement at the base. The equation governing this system is

$$m\ddot{x} + c[\dot{x} - \dot{f}(t)] + k[x - f(t)] = 0 \tag{9.37}$$

Since the force is given by $f(t) = F\sin(\omega t)$, the system equation becomes

$$m\ddot{x} + c\dot{x} + kx = kf(t) + c\dot{f}(t) \tag{9.38}$$

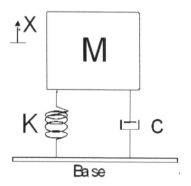

Figure 9.25: Mass-spring-damper system.

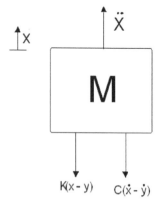

Figure 9.26: Effect of harmonic movement at the base.

To solve this equation, we rewrite the differential equation as a system of first-order linear differential equations. Since

$$f(t) = F\sin(\omega t)$$
$$\dot{f}(t) = F\omega\cos(\omega t)$$

$$x_1 = x$$
$$x_2 = \dot{x} = \dot{x}_1$$
$$\dot{x}_2 = \ddot{x} = -\frac{k}{m}x - \frac{c}{m}\dot{x} + \frac{c}{m}\dot{f}(t) + \frac{k}{m}f(t)$$

$$= -\frac{k}{m}x_1 - \frac{c}{m}x_2 + \frac{cF\omega}{m}\cos(t) + \frac{kF}{m}\sin(t)$$

In terms of x_1, x_2, we have

$$\begin{cases} \dot{x}_1 = x_2 \\ \\ \dot{x}_2 = -\frac{k}{m}x_1 - \frac{c}{m}x_2 + \frac{cF\omega}{m}\cos(t) + \frac{kF}{m}\sin(t) \end{cases}$$

This system of equations can be programmed in a MATLAB script as

```
function xpunto = eq_dif(t, x)
global k w m wn F c
xpunto(1) = x(2);
xpunto (2) = -(c/m)*x(2)-(k/m)*x(1)+(c*F/m)*cos(w*t)+...
   (F*k/m)*sin(w*t);
xpunto = xpunto';
```

The script to solve the system for $m = 120$ Kg and for $m = 500$ Kg is:

```
% This is file Example9_18.m
% Evaluates the path and the position coordinates.
% Time from 0 to 15 seconds.
close all
clear
clc
global k w m wn F c
time = linspace ( 0, 100, 100 );
% Initial conditions
x0 = [ 0 0 ]';
w = 5.81778; m=120;
k = 40e1; wn = sqrt(k/m); F = 0.1; c = wn*m;
% Calling the function to solve the
% system of differential equations.
[ t , x ] = ode23 ( 'eq_dif', time , x0 );
% Plot of displacement versus time
```

```
subplot(2,2,1)
plot( t , x ( : , 2 ) )
grid on
xlabel ( 'Time')
ylabel ( 'Displacement')
legend ( ' m = 120 Kg ' )
subplot(2,2,2)
plot( t , x ( : , 1) )
xlabel ( 'Time')
ylabel ( 'Velocity')
grid on
m = 500; k = 40e1; wn = sqrt(k/m); c = wn*m;
[ t , x ] = ode23 ( 'eq_dif', time , x0 );
subplot(2,2,3)
plot( t , x ( : , 2 ) )
legend( ' m = 500 Kg')
grid on
xlabel ( 'time')
ylabel ( 'Displacement')
subplot(2,2,4)  plot( t , x ( : , 1) )
xlabel ( 'time')
ylabel ( 'Velocity')
grid on
```

The plots produced are shown in Figure 9.27. The two plots at the left are the displacement and the ones to the right are the velocity for the two different values of mass, M. We readily see that when the mass is large, the displacement is small.

Example 9.19 Response of a structure to an impulse

To test the structure of Figure 9.28, we apply an impulse with a hammer. The magnitude of the impulse is F. If the system is underdamped the differential equation that controls its behavior is

$$m\ddot{x} + cx + k\dot{x} = 0 \qquad (9.39)$$

The initial conditions can be established by considering that for $t<0$ we have that $x(0) = \dot{x}(0) = 0$. But at $t = 0^+$ we apply the impulse with a magnitude F. Since the momentum is conserved, we then have that

$$\text{Impulse} * F = F\delta(t) = m\dot{x}(0^-) + m\dot{x}(0^+) = m\dot{x}(0^+) \qquad (9.40)$$

Thus,

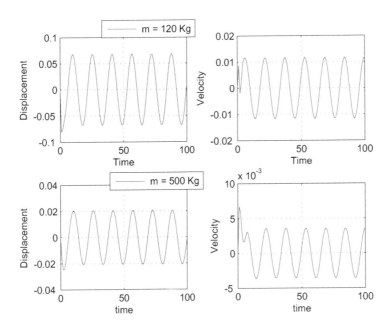

Figure 9.27: Displacement and velocity plots.

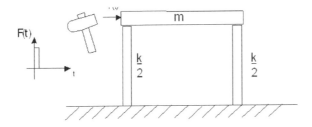

Figure 9.28: Structure excited by an impulse.

$$\dot{x}(0^+) = \frac{F}{m} \qquad (9.41)$$

The displacement x at $t = 0^+$ has the same value it has at $x = 0^-$, that is,

$$x(0^-) = x(0^+) = 0$$

The differential equation can be rewritten as a set of linear first-order differential equations as:

$$\begin{cases} \dot{x}_1 = x_2 \\[2mm] \dot{x}_2 = -\frac{k}{m}x_1 - \frac{c}{m}x_2 \end{cases}$$

This system of equations can be programmed in a MATLAB function as:

```
function xpunto = eq_dif2( t, x )
global m k c f
xpunto(1) = x(2) ; xpunto (2 ) = -(c/m)*x(2)-(k/m)*x(1);
xpunto = xpunto';
```

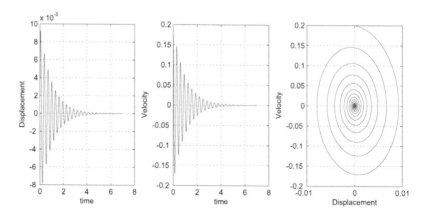

Figure 9.29: Plots of displacement and velocity vs. time and velocity vs. time.

For values $m = 5$ Kg, $k = 2000$ N/m, $c = 10$ N s /m and $F = 10$ N, the following script plots the displacement, the velocity, and a state plot of velocity vs. displacement. The plots are shown in Figure 9.29.

```
% This is file Example9_19.m
% Evaluates the displacement and velocity of a structure
% under an impulse.
```

```
clear
clc
close all
global m k c f
% time from 0 to 7 seconds
time = linspace ( 0, 7, 1000 );
% initial conditions x1=0, x2=F/m.
x0 = [ 0 .2 ]';
% Data
m=5;  k=2e3;
c=10;
% Calling the function to solve
% the system of differential equations
[ t , x ] = ode23 ( 'eq_dif2', time , x0 );
% Plot of displacement
subplot(1,3,1)
plot( t , x ( : , 1 ) )
grid on
xlabel ( 'time')
ylabel ( 'Displacement')
% Plot of velocity
subplot(1,3,2)
plot( t , x ( : , 2) )
grid on
xlabel ( 'time')
ylabel ( 'Velocity')
% Plot of the path
subplot(1,3,3)
plot ( x (:,1), x(:,2))
xlabel ( 'Displacement')
ylabel ( 'Velocity')
grid on
```

9.9   Concluding Remarks

We have presented examples showing how MATLAB can be applied to the solution of many kinds of problems in engineering. The examples presented here are only a sample of many applications. Any problem that can be stated in analytical form or in a numerical one can be solved using the many functions and tools provided by MATLAB.

9.10   References

[1] R.B. Bird, et al. Transport Phenomena, J. Wiley and Sons, NY, 1960.

[2] A. Constantinides and N. Mostoufi, Numerical Methods for Chemical Engineers with MATLAB Applications, Prentice Hall PTR, Upper Saddle River, NJ, 1999.

[3] H. Moskowitz and G. P. Wright, Operations Research, Prentice-Hall Inc, Upper Saddle River, NJ, 1982.

[4] L. Blank and A. Tasquin, Engineering Economy, 5th Ed., McGraw-Hill Book Co., New York, 2004.

[5] M.E. Sosa-Morales, R. Orzuna-Espiritu, and J. Velez-Ruiz, Mass, Thermal and quality aspects of deep-fat frying pork meat, Journal of Food Engineering, Vol. 77, pp. 731-738, 2006.

Chapter 10

MATLAB Applications in Physics

The relationship between physics and mathematics is very strong. This importance has been recognized always and thus the Greeks tried to describe with mathematics all physical phenomena they observed. Nowadays, every physics branch is described mathematically to be understood by peers and students. This chapter presents examples showing how MATLAB can be used to solve physics problems, taking advantage of the m-language and its graphics capabilities to display results. The examples presented in this chapter are on kinematics, dynamics, electromagnetism, optics, astronomy, and modern physics.

10.1 Examples in Kinematics

The examples in this section cover topics related to a course in College Physics. The mathematics required here are as simple as solving an equation, differentiating, integrating, and solving differential equations. However, MATLAB provides very easy-to-use tools for displaying results in a numerical or graphical way. Our examples in this section include constant acceleration, free-fall, parabolic throw, and harmonic motion.

Example 10.1 Constant acceleration

The equations of motion for constant acceleration are given by the following equations[1]:

$$x = x_0 + v_0 t + \frac{1}{2} a t^2 \tag{10.1}$$

$$v^2 = v_0^2 + 2a(x - x_0) \tag{10.2}$$

$$v = v_0 + at \qquad\qquad (10.3)$$

$$x = x_0 + (v_0 + v)t \qquad\qquad (10.4)$$

where x_0 and v_0 are the initial position and initial velocity, respectively.

Now, a motorist is driving at a constant velocity of 72 km/h at a 60 km/h road. A police officer who is standing next to the road spots the speeding car and at the moment the car passes besides it, the officer starts pursuing it with a constant acceleration of 4 m/s². We wish to know how long it will take the officer to catch up with the speeding car?

To solve this problem we have to find the time where both vehicles are together in the same position, that is, they have travelled the same distance. Since both vehicles have a constant acceleration, we can use the equations displayed above. We consider that the officer is at the position $x = 0$. If x_f and t_f are the position and time when the vehicles have travelled the same distance, then using Equation 10.1 we have for the speeding car, which has an acceleration of 0 m/s²

$$x_f = 0 + 72t + \frac{1}{2}0t^2$$

and for the police officer, which has no initial velocity and a constant acceleration of 4 m/s²,

$$x_f = 0 + 0t + \frac{1}{2}4t^2$$

These two simultaneous equations can be solved in MATLAB using the command solve, but first we have to convert the speeding car velocity to m/s. Since 1 km = 1000 m and 1 hour = 3600 s the velocity is then

$$v_{car} = \frac{72,000}{3600}\text{m/s} = 20\text{m/s}$$

The equations are then

$$x_f = 0 + 20t + \frac{1}{2}0t^2$$

$$x_f = 0 + 0t + \frac{1}{2}4t^2$$

The instruction in MATLAB is then

$$[\text{t, x}] = \text{solve('x} = 20\text{*t', 'x} = 4\text{*t\textasciicircum2/2')}$$

which in MATLAB produces

[t, x] = solve('x = 20*t', 'x = 4*t^2/2')

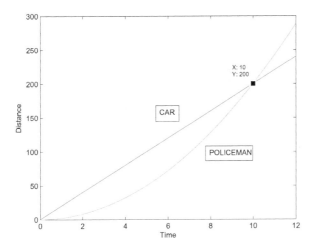

Figure 10.1: Distance travelled by both vehicles.

t =
0
10

x=
0
200

We can see that there are two solutions and one of them is the trivial solution, that is, for t = 0 we get x = 0. The other solution gives t = 10 s and x = 200 m. Then, it takes 200 m for the police officer to catch up the speeding car. The time spent in the pursuit is 10 s. We now can make a plot of the distance travelled by both vehicles. We can do this with

```
t = 0: 0.01: 12;
x_car = 20*t;
x_police = 4*t^2/2;
plot( t, x_car, t, x_police)
xlabel('Time')
ylabel('Distance')
```

Now we plot the velocity for both vehicles using Equation 10.2. For the speeding car the velocity is constant at 72 km/h. For the police officer we have the final speed is

$$v_{police} = \sqrt{v_0^2 + 2a(x - x_0)} = \sqrt{0 + 2*4*200} = 40 \text{ m/s}$$

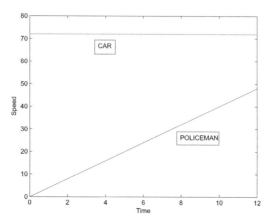

Figure 10.2: Plot of speed vs. time.

A plot of speed for both vehicles can be plotted with

```
t = 0: 0.01: 12;
x_police = 4*t.^2/2;
v_police = sqrt(2*4*x_police);
v_car = 72;
plot(t, v_police, t, v_car)
xlabel('Time')
ylabel('Speed')
```

The resulting plot is shown in Figure 10.2.

Example 10.2 Variable acceleration

Now suppose that the police car accelerates with an acceleration given by

$$a = t \ \text{m/s}^3 \tag{10.5}$$

Since the acceleration is the derivative of the velocity, then we have that

$$v_f = v_0 + \int a(t)dt = 0 + \int tdt = \frac{t^2}{2} \tag{10.6}$$

Also, the distance travelled by the police car is the integral of the velocity. Then,

$$x = x + \int v(t)dt = 0 + \int \frac{t^2}{2}dt = \frac{t^3}{6} \tag{10.7}$$

For the speeding car the position is given as before by

$$x = 20t \tag{10.8}$$

We can solve again for position and time with MATLAB with

[t, x] = solve('x = 20*t', 'x = 4*t^3/6')
t =
0
30^(1/2)
-30^(1/2)

x =
0
20*30^(1/2)
-20*30^(1/2)

MATLAB gives a symbolic solution that we can convert to a real number with

double([t, x])

ans =
0   0
5.4772   109.5445
-5.4772   -109.5445

We see that there are three solutions. This is because we have a cubic equation. Of the three solutions only the second one is a valid one. (The first solution is the trivial solution and the last one has negative time.) Thus, the time it takes the police car to reach the speeding car is $t = 5.4772$ s. Because the acceleration increases with time, the police car pursues the speeder for a shorter period of time.

Example 10.3 Free-falling ball

A baseball drops from a building roof 40 m high. We wish to know how long it will take to hit the sidewalk and what is the velocity at the time of the impact. This is a constant acceleration movement and the value of acceleration is that due to gravity which has the value $g = 9.8$ m/s². The initial velocity is $v_0 = 0$. Using Equations 10.3 and 10.4 we get the set of simultaneous equations

$$v = 0 - gt \tag{10.9a}$$

$$0 = 40 + vt \tag{10.9b}$$

We can solve this set of equations with

> **[t, v] = solve('v = -9.8*t', 'vt+40 = 0')**
> t =
> -2.02030508910442149828812674887 10
> 2.02030508910442149828812674887 10
> v =
> 19.7989898732233306832236421 38936
> -19.7989898732233306832236421 38936

Again, this is a symbolic result which can be converted to a real one with

> **double([t, v])**
> ans =
> -2.0203   19.7990
> 2.0203   -19.7990

The result with $t < 0$ is not a valid solution and, thus, the solution is

$$t = 2.0203 \text{ sec} \qquad v = \text{-19.799 m/s}$$

that is, the time it takes the ball to hit the floor is 2.0203 s and its speed is 19.799 m/s.

Example 10.4 Parabolic throw

A projectile is thrown with an initial velocity of $v_0 = 27$ m/s and with an angle $\theta = 57°$ with respect to the horizontal. Find the maximum distance that the projectile will travel before hitting the ground.
The velocity components are

$$v_{ox} = 27\cos\theta = 14.7053 \quad \text{and} \quad v_{oy} = 27\sin\theta = 22.6441$$

For the path coordinates (x, y), we have that the y component is affected by the gravity acceleration g. The x component, ignoring air resistance, is not affected by gravity. Thus, the problem solution is found if we find the position of the ball when $y = 0$. To solve this problem we define first the acceleration components for the projectile. They are:

$$\ddot{x} = 0$$
$$\ddot{y} = -g$$

They can be written as a vector of first order differential as

$$\dot{x}_1(t) = x(t) \qquad\qquad \text{x-component of the ball position}$$
$$\dot{x}_2(t) = y(t) \qquad\qquad \text{y-component of the ball position}$$
$$\dot{x}_1(t) = \dot{x}(t) = x_3(t) \quad \text{x-component of the ball velocity}$$
$$\dot{x}_2(t) = \dot{y}(t) = x_4(t) \quad \text{y-component of the ball velocity}$$

We can rewrite these equations as

$$\dot{x}_1(t) = x_3(t)$$
$$\dot{x}_2(t) = x_4(t)$$
$$\dot{x}_3(t) = \ddot{x}(t) = 0$$
$$\dot{x}_4(t) = \ddot{y}(t) = -g = -9.8$$

The initial conditions for the variables are

$$x_1(t) = 0$$
$$x_2(t) = 0$$
$$x_3(t) = v_{x0} = 14.7053$$
$$x_4(t) = v_{y0} = 22.6441$$

This set of first-order simultaneous differential equations can be solved if we first write them in an m-function as

```
function x_dot = projectile(t, x)
g = 9.8; % gravitational acceleration.
% Initialization of the vector of derivatives.
x_dot(1) = x(3);
x_dot(2) = x(4);
x_dot(3) = 0;
x_dot(4) = -9.8;
x_dot = x_dot';
```

This set of differential equations can be solved with the following script:

```
% This is file Example_10_4.m
% This script solves the motion equations for the projectile throw
% and plots the path.
dt = 0: 0.001: 5;
[t, x] = ode45('projectile', dt, [0, 0, 14.7053, 22.6441])
for n = 1: 5000;
    if x(n,2) <= 0
        x(n,2) = 0;
    end
end
plot(t, x(:, 2))
xlabel('Time')
ylabel('y-component')
```

The position is plotted in Figure 10.3 and there we see the maximum height and the time when the ball falls to the ground.

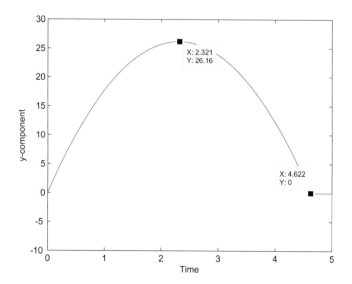

Figure 10.3: Plot of the y-coordinate for the position vs. time.

10.2   Examples in Dynamics

In dynamics, the forces acting on the objects have to be taken into consideration. Our first example is the parabolic throw but this time considering the effect of air resistance. Our second example is the movement of the pendulum, and our last example is a system composed of a mass, a spring, and a damper.

Example 10.5 Parabolic throw with air resistance

We here analyze the parabolic throw but this time we take into consideration air resistance. The air opposes to the movement of the ball with a force that is proportional to the velocity squared as

$$\vec{F} = -cv^2 \frac{\vec{v}}{|\vec{v}|} \tag{10.10}$$

where c is a known constant. The equations of motion are now

$$m\ddot{x} = F_x = -c\dot{x}\sqrt{\dot{x}^2 + \dot{y}^2} \tag{10.11a}$$

$$m\ddot{y} = -mg + F_y = -mg - c\dot{y}\sqrt{\dot{x}^2 + \dot{y}^2} \tag{10.11b}$$

If we write these equations as a system of first-order ordinary differential equations we have

$$\dot{x}_1(t) = x_3(t)$$
$$\dot{x}_2(t) = x_4(t)$$
$$\dot{x}_3(t) = \ddot{x}(t) = -\frac{c}{m}\dot{x}\sqrt{\dot{x}^2 + \dot{y}^2} = -\frac{c}{m}\dot{x}_3\sqrt{x_3^2 + x_4^2}$$
$$\dot{x}_4(t) = \ddot{y}(t) = -\frac{g}{m} - \frac{c}{m}\dot{x}_4\sqrt{x_3^2 + x_4^2}$$

This system can be programmed in a function as

```
function x_dot = projectile_air(t, x)
g = 9.8; % gravitational acceleration.
c = 0.001; % Constant in the air force equation.
m = 1; % mass of the ball.
% Initialization of the vector of derivatives.
x_dot(1) = x(3);
x_dot(2) = x(4);
x_dot(3) = -(c/m)*x(3)*sqrt(x(3)^2+x(4)^2);
x_dot(4) = -g/m-(c/m)*x(4)*sqrt(x(3)^2+x(4)^2);
x_dot=x_dot';
```

The following file solves the system of differential equations:

```
% This is file Example_10_5.m
% This script solves the motion equations for the projectile throw
% and plots the path when we consider the effect of air resistance.
dt = 0: 0.001: 5;
[t, x] = ode45('projectile_air', dt, [0, 0, 14.7053, 22.6441])
for n = 1: 5000;
    if x(n,2) <= 0
        x(n,2) = 0;
    end
end
plot(t, x(:, 2))
xlabel('Time')
ylabel('y-component')
```

The results of this script are shown in Figure 10.4. We readily see that the distance travelled by the ball has been reduced. The height is also less than the height when there is no air resistance.

Example 10.6 Simple pendulum

The pendulum is one of the most common examples in physics showing a harmonic movement. The pendulum is a mass m attached to a rod of length

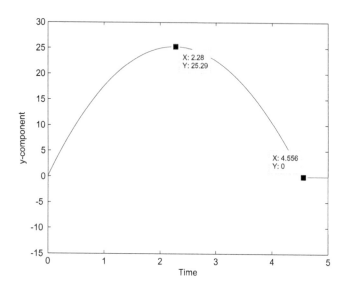

Figure 10.4: Plot of the y-coordinate for the position vs. time when air resistance is taken into account.

L of negligible mass and with a damping coefficient B. The angle θ is the angle between the vertical line and the rod. The equation of motion is

$$F_T = -w \sin \theta - BL\dot{\theta} \tag{10.12}$$

where F_T is the tangential force acting on the mass m, w is the weight given by $w = mg$, g is the acceleration of gravity. From Newton's second law we have

$$F_T = m\frac{dv}{dt} = mL\ddot{\theta} \tag{10.13}$$

Then, the equation of motion for the pendulum is

$$mL\ddot{\theta} + BL\dot{\theta} + w \sin \theta = 0 \tag{10.14}$$

This differential equation, for small values of θ, can be written as

$$mL\ddot{\theta} + BL\dot{\theta} + w\theta = 0 \tag{10.15}$$

because for small θ we have $\sin \theta \approx \theta$. This differential equation can be solved with dsolve. Assuming that $m = 2$ Kg, $L = 0.6$ m, and $B = 0.08$ Kg/m/s, we have

$$0.6\ddot{\theta} + 0.048\dot{\theta} + 2\theta = 0$$

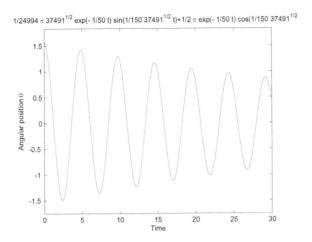

Figure 10.5: Plot of angular position vs. time.

Now, if the initial position of the pendulum is

$$\theta = \pi/2 \qquad \dot{\theta} = 0$$

the instruction dsolve is then

x = dsolve('1.2*D2x+0.048*Dx+2*x=0','x(0)=pi/2','Dx(0)=0')

which gives the solution

**x = 1/24994*pi*37491^(1/2)*exp(-1/50*t)*sin(1/150*37491^(1/2)*t)....
+1/2*pi*exp(-1/50*t)*cos(1/150*37491^(1/2)*t)**

We now plot this solution with the instruction ezplot as

```
ezplot(x,[0,10])
xlabel('Time')
ylabel('Angular position \theta')
```

The resulting plot is shown in Figure 10.5. We see how the damping factor makes the oscillation amplitude decrease.

Example 10.7 Simple pendulum for large θ

Now we analyze the behavior of the pendulum when the oscillation is large, that is, the angle θ is large so the approximation $\sin\theta \not\approx \theta$ is not valid. Thus, we use Equation 10.14 repeated here for convenience:

$$mL\ddot{\theta} + BL\dot{\theta} + w\sin\theta = 0 \qquad (10.16)$$

This equation cannot be solved with the instruction dsolve because it cannot find a closed-form solution. We have to use numerical methods to find a solution. We first have to obtain a set of linear differential equations. We define $x_1(t) = \theta$ and $x_2(t) = \dot{\theta}$. Now, Equation 10.16 can be written as

$$\dot{x}_1(t) = x(2) \qquad (10.17a)$$

$$\dot{x}_2(t) = -\frac{B}{m}x_2(t) - \frac{w}{mL}\sin[x_1(t)] \qquad (10.17b)$$

These equations are described in the file pendulum.m as:

```
function xdot = pendulum(t, x)
g = 9.8; % gravitational acceleration
w = 2; % pendulum weight
L =0.6; % Rod lenght
B = 0.08; % Damping factor
m = w/g; % Mass of the pendulum
% Initialization of the vector of derivatives.
xdot(1) = x(2);
xdot(2) = -B/m*x(2)-w/(m*L)*sin(x(1));
xdot = xdot';
```

Using the file Example_10_7.m, we obtain the plots for the solutions shown in Figure 10.6. This is a better approximation for the pendulum oscillations because we used the exact equation of motion.

```
% This is file Example_10_7.m
t0 = 0 ; % initial time.
tf = 10;
x0 = [ pi/2 0]; % initial conditions.
[t, x] = ode45('pendulum', [t0, tf], x0);
plot (t, x);
legend( '\theta','angular velocity') xlabel('time')
```

Example 10.8 Parachute diving

Let us suppose a parachuter jumps in free fall from an airplane at $t= 0$ s. from a height h and opens the parachute at time t_0. From the initial jump up to time t_0 he/she goes in free fall with acceleration g. The velocity in this time interval is $v(t) = gt$ and the elevation is $y(t) = h - gt^2/2$. At time t_0 the parachute opens. At this time the initial conditions are

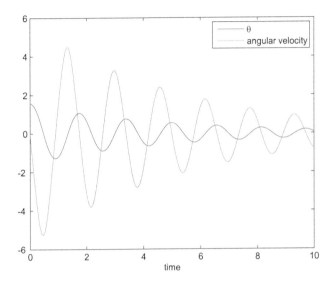

Figure 10.6: Plot of angular position and speed vs. time for exact equations.

$$v(t_0) = gt_0 \tag{10.18a}$$

$$y(t_0) = h - g\frac{t_0^2}{2} \tag{10.18b}$$

The parachute must open in a time t such that $y(t) > 0$. That is, from Equation 10.18(b) we must have that

$$h > g\frac{t_0^2}{2} \tag{10.19}$$

In other words, we must have $t_0 < 2h/g$.

When the parachute is opened, the speed decreases due to air resistance which becomes evident as a drag force proportional to the velocity squared and the equation of motion becomes

$$m\frac{dv}{dt} = -mg + kv^2 \tag{10.20}$$

where m is the parachuter mass and k is the drag coefficient (at sea level has the value 1.29 kg/m^3) which is a function of the air density ρ and the parachute cross-sectional area A. It can be approximated as

$$k = 0.04\rho A \tag{10.21}$$

When the drag force equates the gravity, the parachuter reaches the maximum velocity because then we have a vanishing acceleration. The equation of motion becomes

$$0 = -mg + kv^2$$

and the final velocity is

$$v_f = \sqrt{\frac{mg}{k}}$$

To solve the equation of motion we write it as

$$\frac{dv}{dt} = -g + \frac{k}{m}v^2$$

and we describe it in the function:

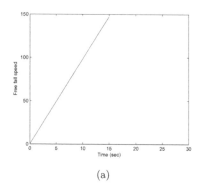

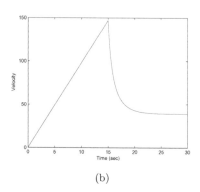

(a)   (b)

Figure 10.7: Parachute diving. (a) Free fall, (b) Parachute open.

```
function equation = parachuter(t, v)
rho = 1.29; % air density.
h = 2000; % height.
A = 10; % parachute cross-sectional area.
g = 9.8; % gravity acceleration.
m = 80; % parachuter mass.
t0 = 15; % time when the parachute is opened.
ti = 0; % initial time.
k = 0.04*rho*A; % Drag coefficient.
equation = - (-g+(k/m)*v.*v);
```

We run this function with the file:

```
% This is file Example_10_8.m
t0 = 15;
t1 = linspace( 0, t0, 41);
```

```
tf = 30;
dt2 = linspace( t0, tf);
g = 9.8;
v1 = g*t1;
plot(t1, v1)
xlabel('Time (sec)')
ylabel('Free fall speed')
axis([0 30 0 150])
fprintf('Press enter to continue')
pause
hold on
v0 = v1(41);
[t, v] = ode45('parachuter', [t0, tf], v0);
plot(t, v)
ylabel('Velocity')
```

The results are shown in Figure 10.8 where we see the free fall velocity from the beginning of the jump to the time when the parachute is opened. Then we see how the speed begins to decrease until it reaches a constant value.

10.3 Applications in Astronomy

Astronomy is an area where mathematics plays an important role and thus, MATLAB can find many applications. In this section we show how to plot the orbit of a planet around the Sun and how Mercury moves around the Sun as seen from Earth.

Example 10.9 Orbit of a planet around the Sun

To plot a planet's orbit around the Sun, we suppose that it has a mass m such that if M is the Sun's mass we have that $M \gg m$. The force between the Sun and the planet is given by Newton's law of universal gravitation:

$$F = G\frac{Mm}{r^2} \tag{10.22}$$

The components of this force are

$$F_x = F\frac{x}{r} = G\frac{Mmx}{r^3} \tag{10.23a}$$

$$F_y = F\frac{y}{r} = G\frac{Mmy}{r^3} \tag{10.23b}$$

From Newton's second law

$$m\frac{dv_x}{dt} = -G\frac{Mmx}{r^3} \tag{10.24a}$$

$$m\frac{dv_y}{dt} = -G\frac{Mmy}{r^3} \tag{10.24b}$$

which can be rewritten as

$$\frac{dv_x}{dt} = -G\frac{Mx}{r^3} \tag{10.25a}$$

$$\frac{dv_y}{dt} = -G\frac{My}{r^3} \tag{10.25b}$$

Since we are only interested in finding the planet's orbit, we set $GM = 1$, we assume that the initial position is $(x, y) = (0.5, 0)$, and that the initial velocity has the components

$$v_x(0) = 0 \qquad v_y(0) = 1.63$$

If we set $v_x = \dot{x}$ and $v_y = \dot{y}$, the equations of motion are now

$$\ddot{x} = -\frac{x}{(x^2 + y^2)^{3/2}} \tag{10.26a}$$

$$\ddot{y} = -\frac{y}{(x^2 + y^2)^{3/2}} \tag{10.26b}$$

These equations can be solved by MATLAB if we write them as a set of linear ordinary differential equations as

$$\begin{aligned} x_1 &= x \\ x_2 &= \dot{x} = \dot{x}_1 = v_x \\ x_3 &= y \\ x_4 &= \dot{y} = \dot{x}_3 = v_y \end{aligned} \tag{10.27}$$

and finally, the system is

$$\dot{x}_1 = x_2 \tag{10.28a}$$

$$\dot{x}_2 = -\frac{x_1}{(x_1^2 + x_3^2)^{3/2}} \tag{10.28b}$$

$$\dot{x}_3 = x_4 \tag{10.28c}$$

$$\dot{x}_4 = -\frac{x_3}{(x_1^2 + x_3^2)^{3/2}} \tag{10.28d}$$

This system can be solved in MATLAB using the function:

```
function xdot = planet(t, x)
xdot(1) = x(2);
xdot(2) =-x(1)/(x(1)^2+x(3)^2)^(3/2);
xdot(3) = x(4);
xdot(4) =-x(3)/(x(1)^2+x(3)^2)^(3/2);
xdot = xdot';
```

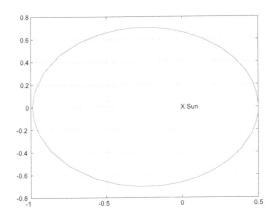

Figure 10.8: Orbit of a planet around the Sun.

We run this function with the m-file:

```
% This is file Example_10_9.m
%
% This file evaluates the trajectory of a planet around the Sun.
x0 = [0.5, 0, 0, 1.63];
x = [0 0 0 0];
[t, x] = ode23('planet', [0, 4], x0);
plot(x(:, 1), x(:, 3))
grid on
text( -0.01, 0, 'X Sun')
```

The resulting trajectory is shown in Figure 10.8. We readily see that the orbit is elliptical.

Example 10.10 Mercury's orbit

Mercury has a very peculiar orbit around the Sun. It is described by the equations

$$x(t) = 93 \cos t + 36 \cos 4.15t$$
$$y(t) = 93 \sin t + 36 \sin 4.15t \tag{10.29}$$

These equations can be plotted with the following m-file:

```
% This is file Example_10_10.m
% This file plots Mercury's orbit around the Sun as seen from the
% Earth.
% Time:
```

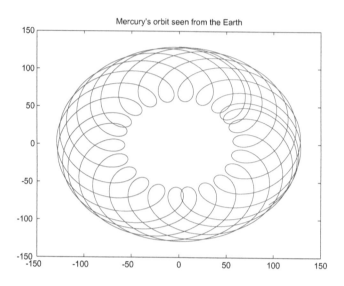

Figure 10.9: Mercury's orbit as seen from Earth.

```
t = [0: pi/360: 2*pi*22/3];
x = 93*cos(t)+36*cos(4.15*t); % x-coordinate.
y = 93*sin(t)+36*sin(4.15*t); % y-coordinate.
axis('square') % This makes a square plot.
plot(y, x)
axis('normal')
title('Mercury''s orbit seen from the Earth')
```

10.4   Applications in Electricity and Magnetism

The examples in this section plot the electrical field of a point charge and the magnetic field produced by a current flowing in a wire.

Example 10.11 Electric field produced by a charge point

The electric field due to a point charge is given by

$$\vec{E}(r) = \frac{1}{4\pi\epsilon_0} \frac{Q}{r^2} \tag{10.30}$$

where r is the distance from the point charge to the observation point. If we only wish to plot the field lines, we can normalize Equation 10.30 with respect to $Q/4\pi\epsilon_0$. Then, Equation 10.30 becomes

$$\vec{E_n}(r) = \frac{\vec{E}(r)}{\frac{Q}{4\pi\epsilon_0}} = \frac{1}{r^2}\hat{r} = \frac{1}{\sqrt{x^2+y^2}}\hat{r} \tag{10.31}$$

where (x, y) are the coordinates of point r. The unit vector $\hat{r}$ is given by

$$\hat{r} = \cos\theta + \sin\theta \tag{10.32}$$

where

$$\theta = \tan^{-1}\frac{y}{x}$$

and the normalized electric field components are given by

$$\vec{E_n} = \left(\frac{1}{\sqrt{x^2+y^2}}\cos\theta, \frac{1}{\sqrt{x^2+y^2}}\sin\theta\right) \tag{10.33}$$

To plot the field lines we first define a grid with x, y values:

$$[\text{x, y}] = \text{meshgrid}(-10\text{: } 2\text{: } 10);$$

Then, the instruction

$$\text{quiver}(\text{x, y, Ex, Ey})$$

plots the field lines as arrows. Finally, we indicate the point charge position with

$$\text{text}(-0.01, 0, \text{'O'})$$

The following m-file plots the electric field lines:

```
% This is file Example_10_11.m
% This file plots the electric field lines
% of a point charge.
[x, y] = meshgrid(-5: 1.25: 5);
%
% Electric field equation
E = 1./(sqrt(x.2+y.2));
% Unit vectors:
[unit_x] = cos(atan2(y, x));
[unit_y] = sin(atan2(y, x));
% Electric field components:
Ex = E.*unit_x;
Ey = E.*unit_y;
% plot
quiver(x, y, Ex, Ey)
text(-0.1, 0, 'O')
```

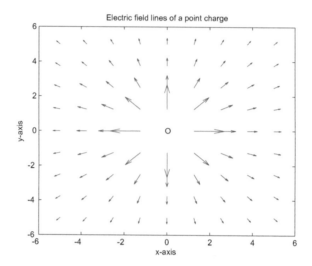

Figure 10.10: Electric field lines.

```
xlabel('x-axis')
ylabel('y-axis')
title('Electric field lines of a point charge')
```

Figure 10.10 shows the field lines due to the electric point charge.

Example 10.12 Magnetic field lines

An electric current flowing in a wire generates a magnetic field. This discovery [2] led to a great deal of activity in electrodynamic research. Basically, if there is a current I flowing in a wire, a magnetic field is produced and it is given by

$$\overrightarrow{\mathbf{B}}(r) = \frac{\mu_0}{2\pi r^2} \overrightarrow{I} \times \overrightarrow{r} \tag{10.34}$$

The magnetic field lines are concentric circles centered in the wire. In the case we have a wire coming out of the paper, we have magnetic field lines in planes parallel to the paper. If the x, y plane is on the paper and the z-axis is coming out of the paper, then the current is

$$\overrightarrow{I} = I\hat{\mathbf{k}} \tag{10.35}$$

where $\hat{\mathbf{k}}$ is the z-axis unit vector. Since we are interested in the field lines we can normalize as we did in the previous example. Then

$$\overrightarrow{\mathbf{B}_n}(r) = \frac{1}{r^2} \overrightarrow{I} \times \overrightarrow{r} \tag{10.36}$$

To plot the magnetic field lines we first need to make a grid with:

$$[\mathsf{x},\ \mathsf{y}] = \mathsf{meshgrid}(\text{-}20\!: 4\!: 20);$$

and then we calculate the vector product given by Equation 10.36

$$\overrightarrow{\mathbf{B}_n}(r) = \frac{1}{r^2}\overrightarrow{I} \times \overrightarrow{r} = \frac{1}{r^2}\begin{vmatrix} \hat{\mathbf{i}} & \hat{\mathbf{j}} & \hat{\mathbf{k}} \\ 0 & 0 & I \\ x & y & 0 \end{vmatrix} = I(-y\hat{\mathbf{i}} + x\hat{\mathbf{j}})$$

The magnetic field components are

$$B_x = -\frac{y}{x^2 + y^2}$$

$$B_y = \frac{x}{x^2 + y^2}$$

use

$$\mathsf{quiver}(\mathsf{x},\ \mathsf{y},\ \mathsf{Bx},\ \mathsf{By})$$

The following m-file plots the magnetic field lines shown in Figure 10.11:

```
% This is file Example_10_12.m
% It plots the magnetic field lines.
% Values for x,y.
[x,y] = meshgrid(-20: 4: 20);
Bx = -y./(x.2+y.2);
By = x./(x.2+y.2);
quiver(x, y, Bx, By)
title('Magnetic field lines')
```

10.5 Applications in Optics

Almost every phenomenon in optics is described by mathematical equations. In this chapter we plot the diffraction pattern of a two-slit screen. An interference pattern produced by two circular waves is plotted and there we can appreciate maxima and minima in the interference pattern.

Example 10.13 Diffraction pattern of a two-slit screen

A front wave passes by a two-slit screen and produces a diffraction pattern. If the slits have a width a, d is the distance between the slits, and λ is the light wavelength, the light intensity on a screen is given by

$$I(\theta) = I_m \cos^2 \beta \left(\frac{\sin \alpha}{\alpha}\right)^2 \tag{10.37}$$

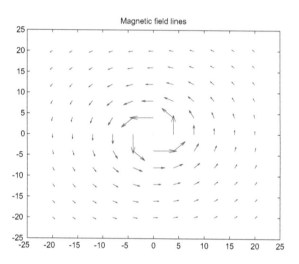

Figure 10.11: Magnetic field lines.

where

$$\beta = \frac{\pi d}{\lambda} \sin \theta \qquad \alpha = \frac{\pi a}{\lambda} \sin \theta$$

The following m-file plots the intensity on a screen when the slit separation is 50λ.

```
% This is file Example_10_13.m
% It plots the diffraction pattern for a two-slit screen.
clear
close all
clc
Theta = [-pi/8:0.001: pi/8];
lambda = 1;
d = 50*lambda; % Slit separation.
a = [ lambda 5*lambda 7*lambda 10*lambda];
for k =1 : 4;
beta = pi*d*sin(Theta)/lambda;
alpha = pi*a(k)*sin(Theta)/lambda;
I = cos(beta).^2.*(sin(alpha)./alpha).^2;
subplot(2, 2, k)
plot(Theta, I)
xlabel('Theta')
ylabel('Intensity')
if k == 1;
title('a = 1')
```

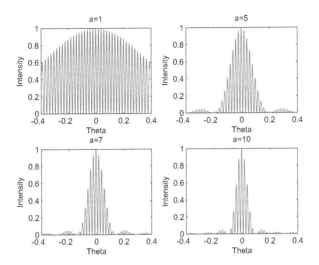

Figure 10.12: Diffraction patterns for a two-slit screen when the slit distance is $d = 50\lambda$.

```
elseif k == 2;
title('a = 5')
elseif k == 3
title('a = 7')
else
title('a = 10')
end
end
```

The results are shown in Figure 10.12. Now, we change d to 6λ and the diffraction pattern changes as shown in Figure 10.13.

Example 10.14 Interference pattern

We can form an interference pattern by drawing two sets of circles. We can do that with the following m-file:

```
% This is File Example_10_14.m
% This file plots two families of circles to observe an
% interference pattern.
close all
N = 256;
t = (0:N)*2*pi/N;% this is the angle that goes from 0 to 2*pi radians.
hold on
```

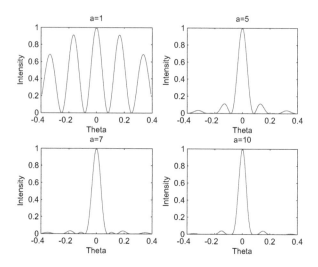

Figure 10.13: Diffraction patterns for a two-slit screen when the slit distance is $d = 6\lambda$.

```
r = 1: 1: 20 % Radii for the circles.
length(r)
for r = 1:length(r)
% Plot of a family of circles.
plot(r(n)*cos(t)-7, r(n)*sin(t), 'LineWidth', 3)
% Plot of the second family of circles.
plot(r(n)*cos(t), r(n)*sin(t), 'LineWidth', 3)
end
```

Figure 10.14 shows the plot where we readily see an interference pattern.

10.6   Applications in Modern Physics

In this section we present two examples in Modern Physics. The first one uses equations from the Special Theory of Relativity and the second example compares energy between Newtonian and modern physics.

Example 10.14 Time dilation

In phenomena taking place at speeds close to the speed of light c we have what is known as time dilation. In this case, if an event takes δt_M seconds in a moving frame, an observer in a reference frame has measured δt_R seconds.

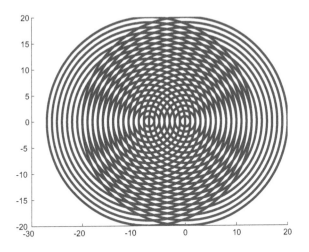

Figure 10.14: Interference pattern.

These time intervals are related by

$$\delta t_R = \frac{\delta t_m}{\sqrt{1 - v^2/c^2}} \tag{10.38}$$

This equation means that a clock at the reference stationary frame sees a longer time than a clock in the moving frame. Now, let us suppose that a twin brother travels at the speed $v = 0.9c$ and makes a round trip to a planet at a distance 2 light-years. How much older is the twin brother when the traveler returns to Earth?

Because we have a round trip, the total distance travelled is 4 light-years, where

$$d = 4 \text{ light-years} = 4 \times 3 \times 10^8 \text{m/s} \times \text{seconds in a year}$$

That is,

$$d = 4 \times 3 \times 10^8 \text{m/s} \times 3.1536 \times 10^6 \text{s} = 3.7843 \times 10^{16} \text{m}$$

For the traveller twin, the time elapsed is

$$\delta t_M = \frac{d}{0.9c} = \frac{3.7843 \times 10^{16}}{0.9 \times 3 \times 10^8} = 1.4016 \times 10^8 \text{s} = 4.4444 \text{ years}$$

On the other hand, for the twin brother on Earth, the time elapsed was

$$\delta t_R = \frac{\delta t_m}{\sqrt{1 - v^2/c^2}} = \frac{4.4444 \text{ years}}{\sqrt{1 - 0.9c/c}} = \frac{4.4444}{\sqrt{0.1}} = 14.0544 \text{ years !!!}$$

Example 10.15 Classic and relativistic energies

The rest energy of an electron is defined by

$$E = mc^2 \tag{10.39}$$

which is the famous Einstein's equation for energy. According to Newtonian physics, when an electron is moving at speed v its kinetic energy is given by

$$K_N = \frac{1}{2}mv^2 \tag{10.40}$$

where m is the electron mass ($m = 9.109 \times 10^{-31}$ g). But according to modern physics, the kinetic energy is given by

$$K_M = \frac{mc^2}{\sqrt{1 - v^2/c^2}} - mc^2 \tag{10.41}$$

A plot of both energies is given in Figure 10.15. This plot is obtained with the following m-file:

```
% This is file Example_10_15.m
% It plots the kinetic energy of an electron moving at speed v.
m = 9.109e-31; % Electron mass.
c = 3e8; % Speed of light.
v1 = logspace(3, 8.95, 1000);
v = logspace(3, 8.47 ,1000);
Kn = m*v1.^2/2; % Newtonian energy.
Km = m*c^2./sqrt(1-v.^2/c^2)-m*c^2; % Relativistic energy.
line([c, c], [0, 4e-13])% Vertical line indicating the speed of light.
plot(v1, Kn, ':', v, Km, 'LineWidth', 3)
title('Comparison between Newtonian and relativistic energies')
xlabel('Speed'), ylabel('Energy')
legend('Relativistic energy','Newtonian energy')
```

The results are shown in Figure 10.15. We readily see how both formulations give the same results for low speeds, but as we approach the speed of light the Newtonian speed yields inaccurate results.

10.7   Concluding Remarks

Physics makes extensive use of mathematics. Thus, many problems are amenable to be solved with a tool as efficient as MATLAB. We have presented problems in kinematics, dynamics, astronomy, optics, and modern physics. Of course, this set is by no means exhaustive but it gives a clear idea of what can be accomplished with MATLAB as a tool to aid in the solution of physics' problems.

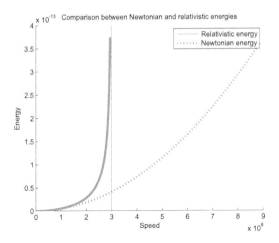

Figure 10.15: Comparison between Newtonian and relativistic energies.

10.8   References

[1] H. D. Young and R. A. Friedman, University Physics, 11th Ed., Addison-Wesley, San Francisco, 2004.

[2] A. Beiser, Concepts of Modern Physics, 6th Ed., McGraw-Hill Book Co., New York, 2002.

[3] R. A. Serway, C. J. Moses, and C. A. Moyer, Modern Physics, Brooks/Cole, Florence, KY, 2004.

[4] F. Jenkins and H. White, Fundamentals of Optics, 4th Ed., McGraw-Hill Book Co., New York, 2001.

Chapter 11

MATLAB Applications in Finance

Finance, similar to engineering, makes extensive use of mathematics. Thus, MATLAB is appropriate to perform computations in these areas. In this chapter we present several financial problems showing how MATLAB can be successfully used to solve them. We start with examples that can be directly solved by MATLAB. Then, we show the use of some functions within the Financial Toolbox and the Financial Derivatives Toolbox. The examples included give the reader an idea of MATLAB's potential to solve problems in Finance. We start this chapter with a brief review of financial concepts and use MATLAB to solve some simple financial problems.

11.1 Simple and Compound Interest

Interest is defined as the manifestation of the value of money in time. We consider three types of interest: simple, compound, and continuously compound [1]. Interest is exclusively calculated from the principal and, as an example, the calculation of simple interest (SI) is done in the following way: If n is the number of periods (usually, we refer to a period as a year, except when otherwise is stated), then for simple interest, after n periods, the final balance is given by

$$\text{Balance\_SI} = \text{Principal} \left[1 + (n) \left(\%\text{Interest Rate} \right) \right] \qquad (11.1)$$

Compound interest (CI) is calculated from the principal plus the total interest over all preceding periods. Thus, it is an interest earned on the interest. It is calculated with the following expression

$$\text{Balance\_CI} = (\text{Principal} + \text{total interest accumulated})(\% \text{ rate of interest}) \quad (11.2)$$

$$= (\text{principal})(1 + \% \text{ rate of interest})^n$$

The above case is known as discrete compounding. If the interest is compounded continuously it is called continuous compounding (CCI). The final result is

$$\text{Balance\_CCI} = (\text{Principal})e^{\% \text{rate\_ of\_ interest} \times \text{time}} \qquad (11.3)$$

Example 11.1 Three types of interest

Let us suppose that a company loans an employee $10,000 with an annual interest rate of 15%. We wish to calculate the employee's total debt after four years, assuming he makes no payments in this period. Thus, Principal=10,000 and rate of interest=0.15. The calculations in MATLAB can be done as follows: (as in previous chapters we use the convention that what the user writes is in bold type and the MATLAB output in regular type). For the simple interest,

Balance_SI = (10000)*(1 + (4)*(.15))

Balance_SI =

16000

For the compound interest,

Balance_CI = (10000)*(1.15)^4

Balance_CI =

1.7490e+004

And for the continuous compound interest

Balance_CCI = 10000*exp(0.15*4)

Balance_CCI =

18221.19

We can compare these results with a bar plot produced by the following script:

```
y = [ BalanceSI, BalanceCI, BalanceCCI ];
bar ( y )
```

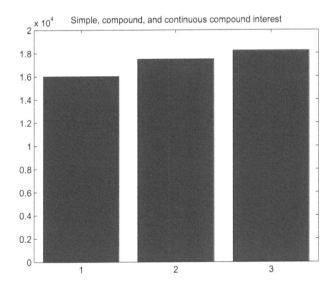

Figure 11.1: Comparison of simple (1), compound (2), and continuous compound (3) interests.

title('Simple, compound, and continuous compound interest')

The results are shown in Figure 11.1. We can also build a plot of the different interest yields as functions of time. Thus, to produce Figure 11.2 we can use the script:

```
n = [0: 0.1: 4];
Balance_SI = (10000)*(1 + (n)*(0.15));
Balance_CI = (10000)*(1.15)^.n;
Balance_CCI = 10000*exp(0.15*n);
plot(n, Balance_SI, '.-', n, Balance_CI, '+', n, Balance_CCI);
title(' Simple, compound, and continuous compound interest')
legend('Simple','Compound', 'Cont. Comp.')
xlabel('years')
ylabel('Balance')
```

Example 11.2 Simple interest

We wish to know the annual simple interest rate so the original investment is doubled in 16 years. In this case the present value of the Principal is C, the

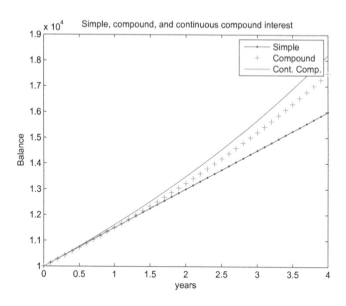

Figure 11.2: Plot of simple, compound, and continuous compound interests.

future value is M = 2 C, and the number of years is n. Then, n = 16 and from Equation (11.1) the equation for the accumulated value is

$$2\ C = C\ (\ 1+\ 16\ R\ )= M \tag{11.4}$$

where R, the annual interest rate is calculated as:

$$R = \frac{M/C\text{-}1}{n} = \frac{2\text{-}1}{16} = 0.0625 = 6.25\% \tag{11.5}$$

The following m-file performs these calculations from the data n, C and M.

```
% This is file for Example 11_2.m
% Evaluates the accumulated value M
% and the number of years n.
close all
clear
clc
C = input ( 'Enter the Present value\n');
M = 2*C;
fprintf ( 'The accumulated value is %5.2fn', M)
n = input ( 'Enter the number of years\n');
R = ( ( 2-1 )/n )*100;
```

```
x='%';
fprintf ( 'The annual interest rate is %5.2f %1sn', R, x)
```

As an example consider the following entry data:

Enter the Present value
2000
The accumulated value is 4000.00
Enter the number of years
16
The annual interest rate is 6.25%

We see that these kinds of computations are very easy to do by using MATLAB.

11.2 Annuities

An annuity is a series of payments, deposits or withdrawals, done in equal periods of time with a compound interest rate. The annuity rent is the payment, deposit or withdrawal that is made periodically and it is denoted by N.

There are several types of annuities but a discussion of them is beyond the scope of the book. Interested readers can consult [1].

It can be shown that the amount of the annuity M in n years with np advanced deposits, where p is the number of deposits per annum at an interest rate r is given by

$$M = \left(1 + \frac{r}{p}\right) \left[\frac{\left(1 + \frac{r}{p}\right)^{np} - 1}{\left(\frac{r}{p}\right)} \right] \qquad (11.6)$$

Example 11.3 Annuities

Now, let us assume that at Laura's birth, her parents make a bank deposit of $6000 and they wish to make monthly payments so that when their daughter is 18 years old they have a sum of $500,000.00. The interest rate is 5% and it is compounded monthly, and we want to know what the monthly required payment should be. The variables in the problem are then

p	12	deposits per annum.
np	18x12	total number of deposits in 18 years.
M	$500,000	amount of the annuity.
N	periodic deposit.	this is the unknown.
r	5% = 0.05	interest rate.

We have then to solve for N in Equation (11.6). We can do this with the following script:

```
syms N
format bank
p = 12;
np = 18*12;
M = 500e3;
r = 0.05; % In the next equation we solve for N
N = solve(M - N*(1+r/p)*(((1+r/p)^(np))-1)/(r/p));
N = double(N)
```

This produces a value for N:

```
N =
    1425.89
```

That is, it is required to make monthly deposits of $1425.89 in order to receive $500,000 after 18 years. We can now generate a plot where the deposit is fixed but the number of deposits is variable. To do this we use the following m-file:

```
% This is file Example11_3.m
% Evaluates the monthly deposit to save
% certain amount if the number of periods is variable.
clear
clc
close all
syms x N
% Datos
p = 12; % number of deposits per year
np = 18*12; %total number of deposits
M = 5e5; %amount of the annuity
i = 0.05; %interest rate
j = 10: 1: np; % plot from 10 to np deposits
for k = 1: length(j);
N = solve (M-x*( 1 + i/p )*(((1+ i/p)^(j(k)))-1)/(i/p));
NN( k )= double(N);
end
plot( j, NN)
grid on
xlabel( 'number of deposits' )
ylabel( 'deposit per period' )
```

As can be seen in Figure 11.3, as the number of deposits increases the amount to be deposited decreases. For example, if Laura's parents make plans for 9

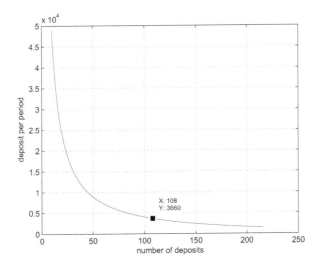

Figure 11.3: Deposits when the number of periods is variable.

years and keep the same goal amount of $500,000, then the number of deposits is just $9 \times 12 = 108$ and the monthly payments increase to $3660.

11.3 Cash Flow

Cash flow refers to the money coming into a person or business from selling services and/or products, saving and earned interest, and the money spent in the production of the products and services.

Example 11.4 Cash flow in savings

Assuming that neither deposits nor withdrawals are made in a 5-year period, it is desired to know what the annual interest rate has to be so the principal doubles at the end of the 5 years.

If the amount deposited initially is Po and the interest rate is r, then we have that the rate of change of the principal P is given by

$$\frac{dP(t)}{dt} = rP \tag{11.7}$$

where the initial condition $P(0) = Po$. This differential equation can be solved with MATLAB using the following instruction:

```
S = dsolve( 'DP -r*P = 0', 'P(0)=Po' )
```

And the solution is

$$S =$$
$$\text{Po*exp(r*t)}$$

Since we are looking for the value of r so the initial investment doubles in 5 years, we have that t = 5 and S = 2Po,

r = solve('2*Po -Po*exp(r*5)')

which gives the symbolic result

$$r =$$
$$\text{1/5*log(2)}$$

That is converted to a numerical value with

r = double(r)
$$r =$$
$$0.1386$$
Or as a percentual value

r = double(r)*100
$$r =$$
$$13.8629$$

The result is then 13.86%. We can check that this result is correct using, from Equation (11.3) from Example 11.1:

$$\text{Balance\_CCI} = (\text{Principal})e^{\%\text{rate\_ of\_ interest} \times \text{time}} \qquad (11.8)$$

Which for a rate of interest r = 13.8629 and t = 5, assuming that the principal at the beginning is \$2000.00, gives,

$$\text{Balance\_CCI} = (2000)*e^{0.138629*5}$$

For MATLAB we have

Balance_CCI = (2000)*exp(0.138629*5)

Balance_CCI =
$$4.0000 \text{ e} + 003$$

This is the double of the initial principal amount, so the interest was correctly calculated.

Example 11.5 Cash flow in investment

On Naomi's birth, her grandfather makes an investment so that when she becomes 18 years old, she can begin college with that money. The annual interest rate is 6.25% compounded continuously. Without an initial deposit, he wishes to know what the monthly deposit should be so the final amount is $2 million. In this case $r = 0.0625$ (6.25%) and d is the cash flow per annum. The differential equation in this case is

$$\frac{dP(t)}{dt} = rP + D \qquad (11.9)$$

And $P(0) = 0$. As in the previous example, we use dsolve to solve the differential equation:

S = dsolve('DP -r*P = d', 'P(0)=0')

which gives the symbolic solution

S =
-d/r+exp(r*t)*d/r

If we wish to rewrite this solution in a more readable form we use pretty as

pretty(S)

$$-d/r + \frac{exp(rt)}{r}$$

Using the values for r, S and t the solution is obtained with:

d = solve('1e6 = -d/0.0625+exp(0.0625*18)*d/0.0625 ')

and

d =
30044.944608783306082858135481384

For the conversion to a numeric quantity,

format bank
d = double(d)
d =
30044.94

Then, the monthly payment must be $30,044.94, since Naomi is born, so that there are $2 million in the account when she turns 18.

Example 11.6 Cash flow 2

Gary opens a bank account with an initial deposit of $2000 with a 5% annual interest rate. Each year he deposits $1000. What is the balance after 10 years for each of the two scenarios below: (a) if the interest is added to the principal in a continuous way and only one $1000 deposit per year is made, and (b) if the interest is added to the principal each quarter and 4 quarterly deposits of $250 are made each year.

Case a.
The data is $r = 0.05$ since the interest rate is 5%, $d = 1000$ which corresponds to the annual deposit, and $Po = 2000$ for the initial deposit or principal. The differential equation is

$$\frac{dP(t)}{dt} = rP + D \tag{11.10}$$

which can be solved with

    S = dsolve('DP -r*P = d', 'P(0)=Po')

To obtain the solution:

    S =
    -d/r+(d+Po*r)/r*exp(r*t)

This is a symbolic solution. To obtain a numerical value we use

    syms Po r t d
    Total = subs(S, r, Po, t, d, 0.05, 2000, 10, 1000)

To get the result:

    Total =
        16271.87

So, the balance after 10 years with one yearly deposit of $1000 at a 5% annual interest rate is $16,271.87.

Case b.
For the quarterly payment case, it can be shown [1] that the balance at the end of n periods is given by

$$P(t) = \left(P_0 + \frac{qd}{r}\right)\left(1 + \frac{r}{q}\right)^n - \frac{qd}{r} \qquad (11.11)$$

where q is the number of periods in a year. Equation 11.11 can be evaluated with MATLAB as:

```
% data
Po = 2000; r = 0.05; t = 10;
d = 250; n = 40; q = 4;
% Calculations
P = ( Po + q*d/r )*( 1 + r/q )^n - q*d/r

P =
    16159.63
```

So, the balance after 10 years with 4 quarterly deposits of $250 at 5% annual interest is $16,159.63. This shows that it is better to make deposits of $1000/year, as the whole yearly amount of $1000.00 starts to earn interest from day 1.

11.4 The Financial Toolbox

MATLAB has developed toolboxes to solve Finance problems. These toolboxes are the Financial Toolbox and the Financial Derivatives Toolbox. Together, they group more than 200 functions that allow us to perform financial computations very easily. In this section, we will only describe a few of them by way of examples so the reader can see the potential of these two toolboxes.

Example 11.7 Bonds with fixed yield

Let us consider a zero coupon bond with a nominal value F, one year maturity and price P. The yield is R where

$$R = \frac{F}{P} \qquad (11.12)$$

The yield rate is

$$r = R - 1 = \frac{F}{P} - 1 \qquad (11.13)$$

which can be written as

$$P = \frac{F}{1 + r} \qquad (11.14)$$

If F and r are fixed, Equation (11.14) can be seen as a discount formula. In the case of cash flow we have a series of periodic payments P_t in discrete times $t = 0, 1, 2, ..., n$. The present value of the cash flow is

$$VP = \sum_{t=0}^{n} \frac{C_t}{(1+r)^t} \tag{11.15}$$

If the frequency of the payments is larger, let us say there are m payments per year, at regular intervals of time, the equation is changed to

$$VP = \sum_{k=0}^{n} \frac{C_t}{(1+r/m)^k} \tag{11.16}$$

where k changes for each period and n is the number of years times the number of payments each year. Now, let us consider the case of a bond with a series of payments $c_t (t = 1, 2,..., n)$. The return yield is defined as the value that makes the present value equal to zero, that is,

$$\sum_{t=0}^{n} \frac{C_t}{(1+\rho)^t} = 0 \tag{11.17}$$

With a change of variable $h = 1/(1+r)$, the equation can be written as

$$\sum_{t=0}^{n} C_t h^t = 0 \tag{11.18}$$

To find the value of ρ we solve this equation in MATLAB with the function roots. If the cash flow is [-100 8 8 8 8 108], a complete procedure is

```
cf = [ -100 8 8 8 8 108 ]; % cash flow.
cf = fliplr ( cf ); % Flip data from left to right
h = roots ( cf ) % Root calculation.
rho = 1./h-1
```

and MATLAB produces

```
h =

-0.8090 + 0.5878i
-0.8090 - 0.5878i
0.3090 + 0.9511i
0.3090 - 0.9511i
0.9259


rho =

-1.8090 - 0.5878i
-1.8090 + 0.5878i
```

    -0.6910 - 0.9511i
    -0.6910 + 0.9511i
    0.0800

Of the five roots, four of them are complex and one is real. The real root gives the rate of return as $\rho = 0.08$. All the above calculations made to find out the rate of return can be easily replaced with the instruction irr in the Financial toolbox as follows:

    cf = [-100 8 8 8 8 108]; % cash flow.
    ro = irr (cf)

    ro =
        0.0800

We readily see the power of the Financial toolbox functions to solve problems in Finance.

Example 11.8 Present value

The instruction pvvar calculates the present value of a series with a discount rate. For example, we can calculate the value of a 5-year bond with a nominal value of 100 and a coupon rate of 8% and 9%. Starting with cf = [0 8 8 8 8 108] we have

    cf = [0 8 8 8 8 108] ;
    pvvar (cf, 0.08) % rate of 8%.

To obtain:

    ans =
        100.0000

If we now repeat with a rate of 9%,

    pvvar (cf, 0.09) % rate of 9%.

we obtain:

    ans =
        96.1103

Note that the cash flow has a zero in the first position because the coupon is received at the end of the first year.

Example 11.9 Quality of the price change

Given a cash flow series happening at times t_1, t_2,..., t_n, the duration of the series is defined as

$$VP = \frac{\sum\limits_{t=0}^{n} PV(t_n)}{PV} \tag{11.19}$$

where PV is the present value of the series and $PV(t_i)$ is the present value of the cash flow c_i at time t_i, $i = 1, 2,..., n$. For the case of a zero coupon bond which is a cash flow, the duration is simply the time to maturity. When we consider a generic bond we use the yield as the discount rate in the computation of the present value to obtain the Macauley duration, assuming m deposits per annum:

$$D = \frac{\sum\limits_{k=1}^{n} \frac{k}{m} \frac{c_k}{(1 + \lambda/m)}}{\sum\limits_{k=1}^{n} \frac{c_k}{(1 + \lambda/m)^k}} \tag{11.20}$$

The derivative of the price with respect to the yield is given by

$$\frac{dP}{d\lambda} = \frac{d}{d\lambda}\left(\sum\limits_{k=1}^{n} \frac{c_k}{(1 + \lambda/m)^k}\right) = -\sum\limits_{k=1}^{n} \frac{k}{m} \frac{c_k}{(1 + \lambda/m)^{k+1}} \tag{11.21}$$

We now define the modified duration as $D_M = D/(1 + \lambda/m)$, so

$$\frac{dP}{d\lambda} = -D_M P \tag{11.22}$$

We see that the modified duration is related with the slope of the curve price-yield. A better approximation is obtained with the convexity defined by

$$C = \frac{1}{P}\frac{d^2P}{d\lambda^2} \tag{11.23}$$

For a bond with m coupons per year is

$$C = \frac{1}{P(1+\lambda/m)} \sum\limits_{k=1}^{n} \frac{k(k+1)}{m^2} \frac{c_k}{(1 + \lambda/m)^k} \tag{11.24}$$

Using the modified duration and the convexity we can write an approximation to P as

$$\delta P \approx D_M P \delta \lambda + \frac{PC}{2} (\delta \lambda)^2 \qquad (11.25)$$

Now let us assume that we have a chain of four cash flows $(10, 7, 9, 12)$ that happen at times $t = 1, 2, 3, 4$. We can calculate the present value of this series with different yields by using the MATLAB function pvvar:

cash_flow = [10 7 9 12]
%Calculate the present value with a rate of 5%.
p1 = pvvar([0, cash_flow], 0.05)

p1 =
    33.5200

% Calculate the present value with a rate of 5.5%.
p2 = pvvar([0, cash_flow], 0.055)

p2 =
    33.1190

p2-p1

ans =
    -0.4010

In this example we have added a 0 in the first position of the cash flow series because the function pvvar assumes that the first cash flow happens in time $t = 0$. We see that an increase in the yield of 0.005 produces a drop in the price of 0.4010. Now we can calculate the duration and convexity with the MATLAB functions cfdur and cfconv:

cash_flow = [10 7 9 12];
[d1 dm] = cfdur (cash_flow , 0.05)

d1 =
    2.5369

dm =
    2.4161

convexity = cfconv (cash_flow, 0.05)

convexity =
    9.4136

First_order = -dm*p1*0.005

First_order =
                    -0.4049

Second_order = -dm*p1*0.005 + 0.5*convexity*p1*0.005^2

Second_order =
                    -0.4010

which is the same result obtained above for p2-p1. We see that for small changes in the yield, the first-order approximation is adequate but, the second-order approximation is practically exact.

Example 11.10 Fixed income bonds

We recall from Chapter 2 the way date and time are formatted in MATLAB. For this we use datestr(today) and datestr(now) to obtain:

datestr(today)

ans =
        12-Feb-2008

datestr(now)

ans =
        12-Feb-2008 11:23:30

In yield calculations we need these functions. To calculate the yield of a bond we can use the MATLAB function bndprice. The format of this function is

[Price, Accrued_Interest] = bndprice(Yield, Coupon rate, ...
        Settlement date, Maturity)

Price is the clean price to which we must add the Yield to obtain the real price (dirty price). For example, for a Yield of 8% and a Coupon rate of 10%, a Settlement date 10-August-2008, and a Maturity date of 31-December-2020, we have:

[Price, Accrued_interest] = bndprice(0.08,0.1,'10-aug-2008','31-dec-2020')

To obtain:

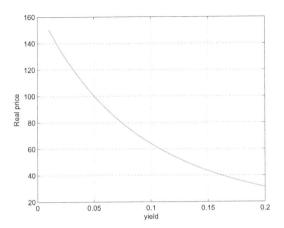

Figure 11.4: Plot of **Real price** vs. **yield.**

Price =
      115.5228
Accrued_interest =
      1.1141

Price + Accrued_interest
ans =
      116.6370

Now, let us consider a vector of yields such as

yield = [0.01 : 0.001 : 0.2];

Repeating the calculation using bndprice with a coupon rate of 5% we obtain
a vector Price and a vector Accrued_interest. With them we can plot the real
price with:

yield = [0.01 : 0.001 : 0.2];
[Price, Accrued_interest] =...
        bndprice(yield, 0.05, '10-aug-2007', '31-dec-2020');
plot (yield, Price + Accrued_interest)
grid on
xlabel ('yield')
ylabel ('Real price')

to produce the plot of Figure 11.4.

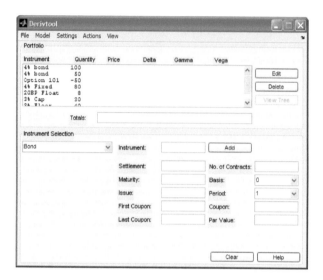

Figure 11.5: **Window for derivtool with some preloaded data.**

From this plot we can readily see that as the yield increases, the Real price decreases.

11.5   The Financial Derivatives Toolbox

In finance, the term derivative refers to a collection of assets falling into three broad categories: options, future contracts, and swaps. A derivative then is a financial instrument whose value depends on the values of other financial instruments such as the price of a stock or of a traded asset [3]. MATLAB has developed the Financial Derivatives toolbox which can be used to analyze individual derivative instruments and portfolios for several types of interest rate based instruments and equity based financial instruments. This toolbox contains a demo graphical user interface (GUI) that illustrates some of the functions that can be analyzed. This demo tool is called derivtool and is run by typing its name in the MATLAB workspace. The GUI for derivtool is shown in Figure 11.5.

As can be seen from the figure, derivtool comes with preloaded data. There is a bond, an option, a fixed, a float, a cap, a floor, and a swap. Figure 11.6 shows the tool with the Option selected.

In the Model menu, the user can choose two models: HJM or Zero Curve. For the Settings we have Initial Curve, Sensitivities, Volatility Model, and Tree Construction. The actions that can be done are Price and Hedge. We can View the Spot Rates and Unit Bond Prices. For example, in the case of the Option 101 we can calculate the Sensitivities by using Settings → Sensitivities. Doing this opens the window of Figure 11.7 and there we select, for example, Delta

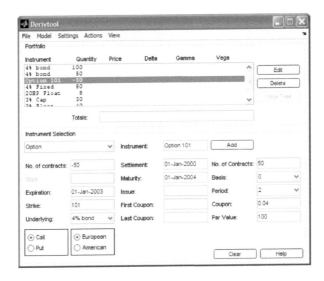

Figure 11.6: Option selected in **derivtool**.

and Gamma. After pressing OK we get the values of these sensitivities for all the instruments in the Portfolio window, as shown in Figure 11.8.

Example 11.11 Hedging with **derivtool**

We can also do hedging with **derivtool**. In Actions we select Hedge. This will do for the 20BP Float and we will hedge with the Floor. After pressing the button Hedge we obtain Figure 11.9 which shows the hedged instrument in the upper right window and the hedging instrument below that window. The lower window provides the sensitivities for both instruments and for the overall portfolio.

11.6 The Black-Scholes Analysis

In 1973, Fischer Black and Myron Scholes published a paper where they derived a differential equation that applies to any derivative that is dependent upon a non-dividend stock [2]. This differential equation can be solved to obtain values for European call and put options on the stock [3]. The so-called Black-Scholes differential equation is

$$\frac{\partial f}{\partial t} + rS \frac{\partial f}{\partial S} + \frac{1}{2}\sigma^2 S^2 \frac{\partial^2 f}{\partial S^2} = rf \tag{11.26}$$

Here, f is the function that satisfies the differential equation. In our case we are looking for the Call option and the Put option. Additionally, S is the stock price, T is the maturity period, r is the risk-free interest rate, and σ is

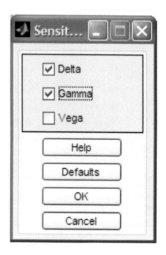

Figure 11.7: Window where sensitivities can be chosen.

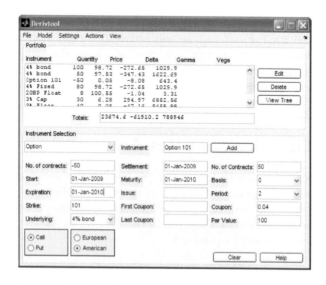

Figure 11.8: Values for **Gamma** and **Delta** sensitivities are shown.

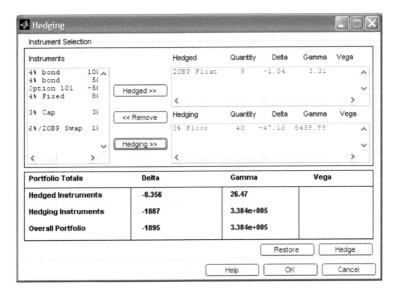

Figure 11.9: Hedging in **derivtool**.

the volatility. The solution depends upon the boundary conditions that are used. A solution to the Black-Scholes differential equation is

For the Call Option

$$c = S_0 N(d_1) Ke^{-rT} N(d_2) \qquad (11.27)$$

and for the Put Option

$$p = Ke^{-rT} N(-d_2) - S_0 N(-d_1) \qquad (11.28)$$

where

$$d_1 = \frac{\ln(\frac{S_0}{K}) + (\frac{r+\sigma^2}{2})T}{\sigma\sqrt{T}} \qquad (11.29)$$

$$d_2 = \frac{\ln(\frac{S_0}{K}) + (\frac{r-\sigma^2}{2})T}{\sigma\sqrt{T}} = d_1 - \sigma\sqrt{T} \qquad (11.30)$$

Here $N(x)$ is the cumulative probability distribution function for a variable that is normally distributed with a mean of zero and a standard deviation of 1, $S0$ is the stock price at time zero, and K is the strike price. The function $N(x)$ is integrated into MATLAB as normcdf(x). A plot of it from $x = -3$ to $x = +3$ is shown in Figure 11.10.

The Financial Derivatives Toolbox has the function blsprice that computes the solution of the Black-Scholes equation. The format for this function is

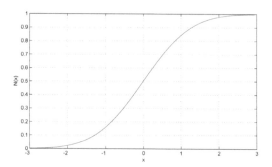

Figure 11.10: Cumulative probability distribution function.

[Call, Put] = blsprice(Price, Strike, Rate, Time, Volatility, Yield)

Example 11.12 European stock price

(This example is taken from [3] John Hull's book "Options, Futures and Other Derivatives," Prentice Hall, Sixth Edition, 2006, Example 12.7, with permission of the author.)

We wish to consider the situation in 6 months from the expiration of an option valued at \$42, the exercise price is \$40, the risk-free interest rate is 10% per annum, and the volatility is 20% per annum. The variables are then, $S = 42$, $X = 40$, $r = 0.1$, $\sigma = 0.2$, and $T = 0.5$. Running the MATLAB function blsprice

[Call, Put]= blsprice (42, 40, 0.1, 0.5, 0.2)

produces the call and put options as

Call =
        4.7594

Put =
        0.8086

This means that the stock price has to rise by \$2.76 for the purchaser of the call to break even. In a similar way, the stock price has to fall by \$2.81 for the purchaser of the put to break even.

We can plot the Call and Put options for several pair of variables. For example, if we are interested in observing the way the put option varies when the interest rate and time change. We can use a mesh plot. The following

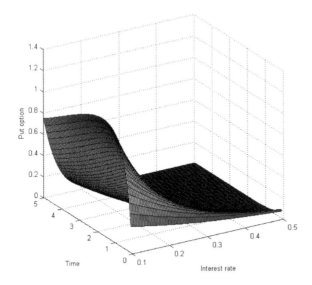

Figure 11.11: Variation of the put option with the interest rate and time.

m-file does the required plot

```
r = 0.1: (0.5 - 0.1)/100: 0.5;
T = 0.1: (5 - 0.1)/100: 5;
for i = 1: length(T)
    for j = 1: length(r)
        [c(i, j), p(i, j)] = blsprice(42, 40, r(j), T(i), 0.2);
    end
end
figure
surf(r, T, p)
xlabel('Interest rate')
ylabel('Time')
zlabel('Put option')
```

In the plot of Figure 11.11 we can observe that the put option begins to increase with time but at about $t = 1.5$ it begins to decrease.

11.6.1 American Options

An American option is an option that can be exercised at any time during the lifespan of the option. Unfortunately, there is no closed form solution to the

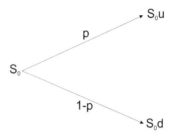

Figure 11.12: Binomial stock price movement.

problem as in the case of European options. Thus, the only way is to proceed with computer methods. The most popular numerical procedures are the use of trees, finite differences methods, and Monte Carlo simulation [3, 6]. The use of trees is a method devised by Cox, Ross, and Rubinstein [4] known as the CRR method.

Binomial trees are studied in many textbooks. The procedure starts by dividing the life of an option into a large number of small time intervals t. In each time interval the stock price moves from its initial value So to one of two new values Sou and Sod. In general, u >1 and d < 1. The movement to Sou is an up movement and the movement to Sod is a down movement. The probability of an up movement is denoted by p and the probability of a down movement is then 1-p. The movement of the stock is shown in Figure 11.12. The parameters u, d, and p are given by

$$u = e^{\sigma\sqrt{\delta t}} \tag{11.31a}$$
$$d = e^{-\sigma\sqrt{\delta t}} \tag{11.31b}$$
$$p = \frac{a-d}{u-d} \tag{11.31c}$$

where a is

$$a = e^{\sigma\delta t} \tag{11.32}$$

The complete tree of stock prices will look as shown in Figure 11.13.

Example 11.13 Binomial tree for an American option

(This example is taken from [3] John Hull's book "Options, Futures and Other Derivatives," Prentice Hall, Sixth Edition, 2006, Example 18.1, with permission of the author.)
Consider a 5-month American put option on a non-dividend-paying stock which costs $50, the strike price is $50, the risk-free interest rate is 10% per annum, and the volatility is 40% per annum. That is, S = 50, K = 50, r =

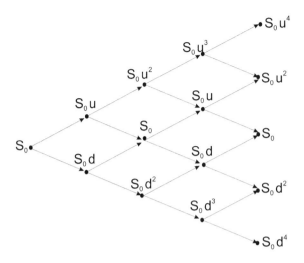

Figure 11.13: Complete binomial tree.

0.1, $\sigma = 0.4$, and $T = 5/12 = 0.4167$. If the time interval is a month, then $t = 0.0833$. The remaining parameters are calculated from Equations (11.31) and (11.32) as

$$u = e^{\sigma\sqrt{\delta t}} = 1.1224$$

$$d = e^{-\sigma\sqrt{\delta t}} = 0.8909$$

$$a = e^{r\delta t} = 1.0084$$

$$p = \frac{a\text{-}d}{u\text{-}d} = 0.5073$$

$$1 - p = 0.4927$$

The MATLAB Financial Toolbox has an integrated function that calculates the values of the nodes of a binomial tree. This function is binprice and has the general form:

[AssetPrice, OptionValue] = binprice(Stock price, Strike price, ...
Rate, Time, Increment, Volatility, Flag, DividendRate, Dividend, ExDiv)

In this function, flag $= 1$ for a call option and flag $= 0$ for a put option. DividendRate, Dividend, and ExDiv are optional parameters and are not used for a non-dividend-paying stock. For a tree with six steps, we have an increment of $(5/12)/6 = 0.0694$. Thus, for a **put option**

[Stockprice,Optionprice]=binprice(50, 50, 0.1,5/12, 0.0694, 0.4,0, 0, 0, 0)

Stockprice =

50.0000	55.5583	61.7344	68.5971	76.2228	84.6961	94.1113
0	44.9978	50.0000	55.5583	61.7344	68.5971	76.2228
0	0	40.4961	44.9978	50.0000	55.5583	61.7344
0	0	0	36.4447	40.4961	44.9978	50.0000
0	0	0	0	32.7986	36.4447	40.4961
0	0	0	0	0	29.5173	32.7986
0	0	0	0	0	0	26.5643

Optionprice=

4.1745	2.0055	0.5882	0	0	0	0
0	6.4610	3.4894	1.2006	0	0	0
0	0	9.6043	5.8893	2.4507	0	0
0	0	0	13.5553	9.5039	5.0022	0
0	0	0	0	17.2014	13.5553	9.5039
0	0	0	0	0	20.4827	17.2014
0	0	0	0	0	0	23.4357

Since we set the **flag** = 0 the option price is for a put option. From the MATLAB output we see that the put option price is 4.1745.

For a **call option** the instruction is

[Stockprice, Optionprice]=binprice(50, 50, 0.1,5/12, 0.0694, 0.4, 1,0,0,0)

And we obtain the value for the **call option** as

Stockprice =

50.0000	55.5583	61.7344	68.5971	76.2228	84.6961	94.1113
0	44.9978	50.0000	55.5583	61.7344	68.5971	76.2228
0	0	40.4961	44.9978	50.0000	55.5583	61.7344
0	0	0	36.4447	40.4961	44.9978	50.0000
0	0	0	0	32.7986	36.4447	40.4961
0	0	0	0	0	29.5173	32.7986
0	0	0	0	0	0	26.5643

Optionprice =

5.9095	9.1275	13.6517	19.6280	26.9124	35.0421	44.1113
0	2.6881	4.6100	7.7067	12.4241	18.9432	26.2228
0	0	0.7521	1.4948	2.9708	5.9043	11.7344
0	0	0	0	0	0	0
0	0	0	0	0	0	0
0	0	0	0	0	0	0
0	0	0	0	0	0	0

That is, the call option is 5.9095.

MATLAB can generate a tree using the function CRRTree available in the Financial Derivatives Toolbox. The general form is

CRRTree = crrtree(StockSpec, RateSpec, TimeSpec)

We now describe each of these arguments.

1.
The argument StockSpec creates the structure of the stock. For our purpose, its form can be simply

Stock_Specification = stockspec(Sigma, AssetPrice)

Where Sigma is the volatility and AssetPrice is the initial stock price. Then,

Stock_Specification = stockspec(0.4, 50)

2.
RateSpec can be generated with intenvset which sets the properties of the interest-rate structure. Additional information on the instruction intenvset is available in Appendix B. For the data of Example 11.13 we set

RateSpec = intenvset('Compounding', -1, 'Rates', 0.1,'StartDates',...
'01-Jan-2003', 'EndDates', '31-May-2003', 'Basis', 3)
which indicates a 10% risk-free interest rate taking a year of 365 days and starting on January 1st, 2003, and ending on May 31st, 2003.

3.
Finally, the time specification is created with crrtimespec whose structure is

TimeSpecification = crrtimespec(ValuationDate, Maturity, NumPeriods)

where ValuationDate is the starting date, Maturity is the ending date, and NumPeriods is the number of periods where we are going to evaluate the tree.

For our example, the number of monthly periods is 5 and

ValuationDate = '1-Jan-2003'; Maturity = '1-June-2003';

Thus,

TimeSpec = crrtimespec(ValuationDate, Maturity, 5)

Running each of these instructions will produce:

Stock_Specification = stockspec(0.4, 50)
Stock_Specification =
FinObj: 'StockSpec'
Sigma: 0.4000
AssetPrice: 50
DividendType: [ ]
DividendAmounts: 0
ExDividendDates: [ ]

RateSpec = intenvset('Compounding', -1, 'Rates', 0.1, 'StartDates',...
'01-Jan2003', 'EndDates', '31-May-2003', 'Basis', 3)

RateSpec =
FinObj: 'RateSpec'
Compounding: 365
Disc: 0.9597
Rates: 0.1000
EndTimes: 150
StartTimes: 0
EndDates: 731732
StartDates: 731582
ValuationDate: 731582
Basis: 3
EndMonthRule: 1

TimeSpec = crrtimespec('1-Jan-2003', '1-June-2003', 5)
TimeSpec =
FinObj: 'BinTimeSpec'
ValuationDate: 731582
Maturity: 731733
NumPeriods: 6
Basis: 0
EndMonthRule: 1
tObs: [0 0.0691 0.1383 0.2074 0.2766 0.3457 0.4148]

dObs: [731582 731607 731632 731657 731682 731707 731733]

Note that dates have been converted to serial date numbers format (see Chapter 2 for date formats). We are now ready to generate the data for the tree with

CRRTree = crrtree(Stock_Specification, RateSpec, TimeSpec)

which produces

```
CRRTree =
FinObj: 'BinStockTree'
Method: 'CRR' StockSpec: [1x1 struct]
TimeSpec: [1x1 struct]
RateSpec: [1x1 struct]
tObs: [0 0.0691 0.1383 0.2074 0.2766 0.3457 0.4148]
dObs: [731582 731607 731632 731657 731682 731707 731733]
STree: {1x7 cell}
UpProbs: [0.5067 0.5067 0.5067 0.5067 0.5067 0.5067]
```

In this output data we see the information of the tree. The tObs variable stores the observation points, dObs contains the dates of the observations in serial date format, and UpProbs contains the probability of an up movement. Now, we can plot the tree with the tree viewer instruction

```
treeviewer(CRRTree)
```

which generates the tree shown in Figure 11.14. In this plot we have selected the nodes for the down movement of the tree. We can compare the results with those obtained from the function binprice and we see that they agree very well, for example, we have the same end value for the tree which is 26.66 and 26.5646 in binprice.

We can also create an instrument portfolio by using

InstSet = instadd('OptStock', {'call'; 'put'}, 50, ValuationDate, Maturity)

The names of the variables we wish to display are

Names = {'Call Option'; 'Put Option'}

So now we update the instrument portfolio with

InstSet = instsetfield(InstSet, 'Index', 1: 2, 'FieldName', {'Name'},...
'Data', Names);

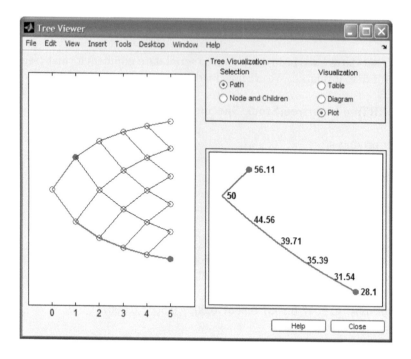

Figure 11.14: Binomial tree.

And create the tree with

[Price, PTree] = crrprice(CRRTree, InstSet)

Price is a vector that contains the call option and the put option values, respectively. Finally, we display the tree with

treeviewer(PTree, Names)

The tree is displayed in Figure 11.15 for the Put Option. This tree looks like the one above but it has a pull-down menu to display the call and put options.

We can compare with the results obtained using the function binprice and there we can see that both results agree.

Example 11.14 CRR tree using a custom GUI

An alternative to the tree construction of the MATLAB toolbox is to write a custom GUI to draw the tree. A very nice example is provided by M. Hoyle [5] from The MathWorks, Inc. He has written the GUI named PlotCRRTree

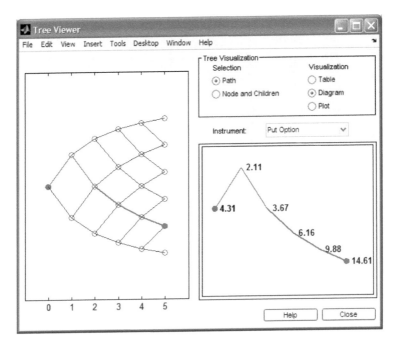

Figure 11.15: Tree for a Put Option with a branch displayed.

that produces a CRR tree. (This GUI is available at The MathWorks Central
File Exchange [5]) The GUI can be run with

PlotCRRTree(S0, K , r, T, sigma, N, flag)

The value of the flag can be true or false depending upon if we wish to display
the option value on the plot. In Figure 11.16 we can see the results for the data
in our example. We can see that there are slide bars to input the data and
by clicking at a node, values for the stock, call and put options are displayed.

11.6.2 Finite Difference Methods

The Black-Scholes partial differential equation can be solved by using finite
differences. In this case the differential equation is converted into a difference
equation that is solved iteratively. We repeat here the Black-Scholes equation

$$\frac{\partial f}{\partial t} + rS\frac{\partial f}{\partial S} + \frac{1}{2}\sigma^2 S^2 \frac{\partial^2 f}{\partial S^2} = rf \tag{11.33}$$

If the life of the option is T and the maximum value of the stock is Smax, we
form a grid by partitioning each of these parameters into N and M equally
divided spaced intervals, respectively. Thus, for T we have N+1 points

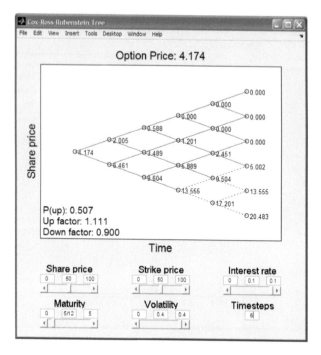

Figure 11.16: Tree for an American option (from M. Hoyle [5]).

$$\delta t, 2\delta t, ..., (N\text{-}1)\delta t, T \tag{11.34a}$$

And for S we have M+1 points

$$0, S, 2\,S, ..., S_{max} \tag{11.34b}$$

The value of S_{max} is chosen in such a way that the current stock price S is a point in the partition. These partitions form a grid with a total of (N+1)×(M+1) points. We denote the value of f at the point (i, j) by f_{ij}. There are two methods for forming the difference equations. We only describe here the implicit difference method. Interested readers can see [3, 6]. The implicit finite difference method makes use of forward difference to approximate $\partial f/\partial t$ by

$$\frac{\partial f}{\partial t} = \frac{f_{i+1,\,j} - f_{i,j}}{\delta t} \tag{11.35}$$

For $\partial f/\partial t$ we use an average of forward and backward differences to obtain

$$\frac{\partial f}{\partial S} = \frac{f_{i,\,j+1} - f_{i,j-1}}{2\delta S} \tag{11.36}$$

An approximation for at the (i, j) point is

$$\frac{\partial^2 f}{\partial S^2} = \frac{f_{i,\,j+1} + f_{i,j-1} - f_{i,j-1}}{\delta S^2} \tag{11.37}$$

The difference equation becomes a difference equation

$$a_j f_{i,\,j-1} + b_j f_{i,j} + c_j f_{i,j+1} = f_{i+1,j} \tag{11.38}$$

where

$$a_j = \tfrac{1}{2}jr\delta t - \tfrac{1}{2}j^2\sigma^2\delta t$$

$$b_j = 1 + \sigma^2 j^2 \delta t + r\delta t \tag{11.39}$$

$$c_j = -\tfrac{1}{2}jr\delta t - \tfrac{1}{2}j^2\sigma^2\delta t$$

A set of boundary conditions is needed and they are the value of the put at time T which is $max(K- ST, 0)$ where ST is the stock price at time T, thus

$$f_{N,j} = max(K-j\delta\,S, 0) \qquad j = 0,\,1,\,2,...,N \tag{11.40}$$

Also the value of the put option when the stock price is zero is K

$$f_{i,0} = K \qquad j = 0,\,1,\,2,...,N \tag{11.41}$$

We assume that the put option is worth zero when $S = Smax$ so that

$$f_{i,M} = K \qquad j = 0,\,1,\,2,...,N \tag{11.42}$$

Now the set of difference equations with the initial conditions can be solved. There is no MATLAB function to solve this set of equations, but a custom m-file can be written for this purpose. A set of m-files has been written by M. Hoyle from the MathWorks, Inc. [5] to solve the differential set of equations. An m-file based on Hoyle's m-files is given in Appendix A. It is named FiniteDiff.m and its format is:

FiniteDiff(Initial price S, Strike price K, Interest rate r, Volatility, Maturity T)

Example 11.15 Evaluating options with the finite differences method

For example, suppose we use the data of Example 11.13 where $S = 50$, $K = 50$, $r = 0.1$ $\sigma = 0.4$, and $T = 5/12 = 0.4167$. Running the m-file FiniteDiff.m we obtain

Finite_diff(50,50,0.1,5/12,0.4)

P_FD =
       4.2424

We can compare this result with the result from the CRR tree method which gives a value of the put option as 4.4885 for a tree with 5 steps but which changes to 4.3046 for a tree with 55 steps. If now we consider the data of Example 11.12 where $S = 42$, $K = 40$, $r = 0.1$, $\sigma = 0.2$, and $T = 6/12 = 0.5$, running the m-file FiniteDiff.m we obtain

Finite_diff(42,40,0.1,6/12,0.2)

P_FD =
       0.8841

Again, this result is very similar to the one obtained by using the CRR method.

11.6.3   Monte Carlo Methods

Monte Carlo simulation is a very useful and important tool in computational finance. It may be used for portfolio management rules, to price options, and for the simulation of hedging strategies, among other uses. The main advantages of these techniques are its generality, ease of use and great flexibility in taking many complicating features into account in the simulation. They are especially useful in treating multidimensional problems where either the closed-form solutions or the CRR and finite difference methods are very difficult to work with. On the other hand, Monte Carlo simulation methods have a computational burden and an increasing number of replications are needed to refine the estimates we are interested in.

We will not discuss the details of the Monte Carlo method but rather we show an example of what can be achieved with this method. The m-function presented here was programmed by Hoyle [5] and it follows the procedure outlined by Longstaff and Schwartz [7]. This procedure starts by generating possible asset price paths, then for a given time step looks at the in the money paths and calculates the discounted payoffs. It uses the least squares regression to model the discounted payoffs. From there, if the early exercise payoff is greater than the model discounted payoff we should exercise the option. The m-function is named LSM_Plot and it is freely available from the MathWorks MATLAB Central site at

http://www.mathworks.com/matlabcentral/fileexchange
/loadFile.do?objectId=16476

The input format for the m-function LSM_Plot is

LSM_Plot(S0 ,K ,r ,T, sigma, N, M)

where

S0	Initial asset price
K	Strike Price
r	Interest rate
T	Time to maturity of option
sigma	Volatility of underlying asset
N	Number of points in time grid to use (minimum is 3, default is 50)
M	Number of points in asset price grid to use (minimum is 3, default is 50)

The m-function LSM_Plot must be within a script with the data to run satisfactorily. An example shows how it works.

Example 11.16 Evaluation of options using Monte Carlo methods

For the same data of Example 11.13 we have that $S = 50$, $K = 50$, $r = 0.1$, $\sigma = 0.4$, and $T = 5/12 = 0.4167$. Additionally, we set $N = 40$ and $M = 100$. We desire to find the data for a call option, thus type $=$ 'True'. Then, we write the following in a script:

```
S = 50; K = 50; r = 0.1; sigma = 0.4;
T = 5/12; N = 40 ; M = 100;
LSM_Plot(50, 50, 0.1, 5/12, 0.4, 40, 10)
```

We obtain the GUI of Figure 11.17 which gives all the runs using a random number generator. This GUI has three windows, one is for the simulation (top left corner), another one below which is useful for a step-by-step simulation and to see the points of exercise. Another window to the right gives the payoff and also makes an interpolation by a quadratic equation to the payoff data, it then compares it with the early exercise payoff and thus we can conclude what is best to do. We see at the low right corner two buttons which can be used to make a step-by-step simulation and to run the complete simulation. Figure 11.18 gives a complete simulation. The MATLAB Command window will display the option value at the end of the simulation which in this example has the value 4.71. Due to the random nature of Monte Carlo methods, this value will change for other runs.

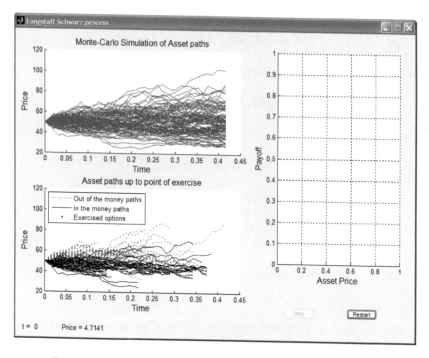

Figure 11.17: GUI for Monte Carlo simulation [5].

11.7   The Greek Letters

MATLAB allows us to calculate the so-called Greek letters. They are useful to traders to manage risks and make them acceptable. MATLAB can evaluate the delta, gamma, lambda, rho, theta, and vega. The instructions to obtain them are the following:

```
[CallDelta, PutDelta] = blsdelta(Price, Strike, Rate, Time, Volatility, Yield)
Gamma = blsgamma(Price, Strike, Rate, Time, Volatility, Yield)
[CallEl, PutEl] = blslambda(Price, Strike, Rate, Time, Volatility, Yield)
[CallRho, PutRho] = blsrho(Price, Strike, Rate, Time, Volatility, Yield)
[CallTheta, PutTheta] = blstheta(Price, Strike, Rate, Time, Volatility, Yield)
Vega = blsvega(Price, Strike, Rate, Time, Volatility, Yield)
```

The parameters in these instructions are the same as in the case of the Black-Scholes case. We show with a few examples the calculation of some of these Greeks.

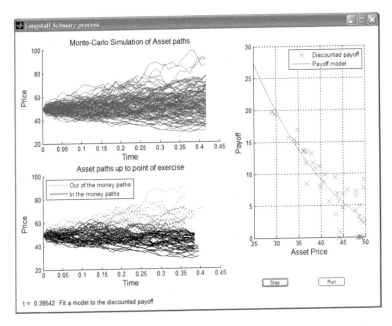

Figure 11.18: A complete Monte Carlo simulation [5].

Example 11.17 Delta evaluation

(This example is taken from [3] in John Hull, "Options, Futures and Other Derivatives," Prentice Hall, Sixth Edition, 2006, Example 14.1, with permission of the author.)

A bank in the USA has sold 6-month put options on 1 million with a strike price of 1.6 and it wishes to make a neutral portfolio. The current exchange rate is 1.62, the risk-free interest rate in the UK is 13% per annum, the risk-free interest rate in the USA is 10% per annum, and the volatility is 15%. The delta for the call and put options can be obtained using Price = 1.62, Strike = 1.6, Rate = 0.10, Time = 0.5, Volatility = 0.15, and Yield = 13%. Thus,

[CallDelta, PutDelta] = blsdelta(1.62, 1.60, 0.10, 0.5, 0.15, 0.13)

which gives

CallDelta =
        0.4793

PutDelta =
        -0.4578

The delta for the put option is -0.4578. It means that when the exchange rate increases by S, the price of the put goes down by 45.78% of S. Thus, the bank has to add a short position of 457800 to the option position to make the delta neutral.

Example 11.18 Theta, Gamma, Vega, Rho, and Lambda evaluation

(This example is taken from Examples 14.3, 14.4, 14.5 and 14.7 in [3] J. Hull, "Options, Futures and Other Derivatives," Prentice Hall, Sixth Edition, 2006, with permission of the author.)

Consider a 4-month put option on a stock index. The current value for the index is 305, the strike price is 300, the yield is 3% per annum, the interest rate is 8% per annum, and the volatility is 25% per annum. That is, Price = 305, Strike = 300, Rate = 0.08, Time = 4/12, Volatility = 0.25, Yield = 3%. Then, the Theta, Gamma, Vega, Rho, and Lambda can be evaluated with

[CallTheta, PutTheta] = blstheta(305, 300, 0.08, 4/12, 0.25, 0.03)

CallTheta =
            -32.4623

PutTheta =
            -18.1528

Gamma = blsgamma(305, 300, 0.08, 4/12, 0.25, 0.03)

Gamma =
            0.0086

Vega = blsvega(305, 300, 0.08, 4/12, 0.25, 0.03)

Vega =
            66.4479

[CallRho, PutRho]= blsrho(305, 300, 0.08, 4/12, 0.25, 0.03)

CallRho =
            54.7893
PutRho =
            -42.5792

[Call_Lambda,Put_Lambda]=blslambda(305, 300, 0.08, 4/12, 0.25, 0.03)

Call_Lambda =

8.3992

Put_Lambda =
        -9.2228

This means that the put option value decrease by Theta/365 per day, that is, 42.5792/365=0.0497 per day. Also, an increase of 1 in the index increases the delta of the option by Gamma=0.0086, approximately. In addition, an increase of 1% in the volatility increases the value of the option by (1%*Vega=) 66.4479, approximately. The value of Rho implies that a 1% change in the interest rate decreases the value of the put option by (0.01*42.6=) 0.426. Finally, the lambda value measures the percent change in the put option price per 1% change in the asset price.

Example 11.19 Plots of the Greeks

We can make plots of the Greeks against the several variables that are involved in their computation. For example, if we wish to plot the Rho against the interest rate and the asset price we can do this with the following script:

```
% plot of Rho
r = 0: 0.001: 0.1;
S = 40:.2:60;
for i = 1: length(r)
for j = 1: length(S)
[CallRho(i, j), PutRho(i,j)] = blsrho(S(i), 50, r(j), 4/12, 0.25, 0.03);
end
end
mesh(r, S, PutRho)
title('Plot of Rho for a Put Option')
xlabel ('Interest rate')
ylabel('Stock Price')
zlabel('Rho for Put Option')
```

The resulting plot is shown in Figure 11.19.
    For a plot of Delta vs. the stock price and the strike price, we can use the following script to obtain Figure 11.20.

```
% plot of Delta
close all
clear all
clc
K = 40: 0.2: 60;
S = 40: 0.2: 60;
```

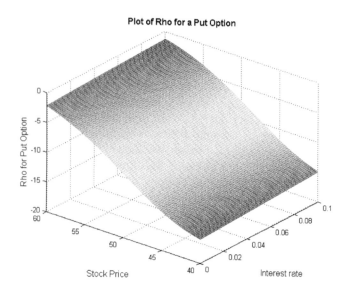

Figure 11.19: Plot of Theta for a call option vs. interest rate and stock price.

```
for i = 1: length(K)
for j = 1: length(S)
[Call_Delta(i, j),Put_Delta(i, j)]=blsdelta(S(j),K(i),0.10, 0.5, 0.15,0.13)
end
end
mesh(r, S, Put_Delta)
title('Plot of Delta for a Put Option')
xlabel ('Strike price')
ylabel('Stock Price')
zlabel('Delta')
```

To plot the Gamma against the stock asset price, we use the following script:

```
% Plot of Gamma
S = 30: 0.2: 50;
for j = 1: length(S)
Gamma(j) = blsgamma(S(j), 40, 0.05, 3/12, 0.15, 0.13);
end
plot(S, Gamma)
title('Plot of Gamma')
xlabel ('Price')
ylabel('Gamma')
```

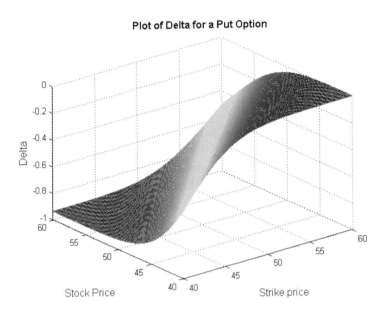

Figure 11.20: Plot of Delta for a put option vs. strike price and stock price.

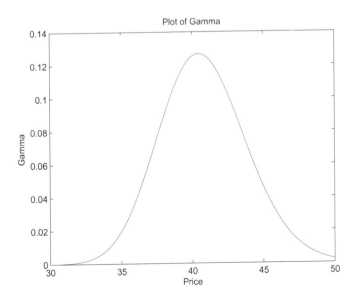

Figure 11.21: Plot of Gamma vs. stock price.

The resulting plot is shown in Figure 11.21. Now if we wish to Theta for a call and put options versus stock price and time we use:

```
clear
close all
clc
T = 0.00001: 0.01: 1.01;
S = 40: 0.2: 60;
for i = 1: length(T)
for j = 1: length(S)
[CallTheta(i,j),PutTheta(i,j)] = blstheta(S(j), 50, T(i), 0.5, 0.15, 0.13);
end
end
mesh(T, S, PutTheta)
title('Plot of Theta for a Put Option')
xlabel ('Time')
ylabel ('Stock Price')
zlabel ('Theta for Put Option')
figure
mesh(T,S,CallTheta)
title('Plot of Theta for a Call Option')
xlabel ('Time')
ylabel ('Stock Price')
zlabel ('Theta for Call Option')
```

The resulting mesh plots are available in Figures 11.22 and 11.23.

11.8   Concluding Remarks

We have presented techniques to solve financial problems using MATLAB. The range of techniques goes from very simple problems, such as interest calculation, to very complicated ones, such as CRR trees, Finite difference and Monte Carlo methods. MATLAB's potential is best appreciated in computing intensive techniques such as the Monte Carlo and sensitivities calculation. The examples provided here are merely representative of the finance functions that MATLAB can compute and, as always, it is up to the reader to take full advantage of MATLAB. Further information on the functions available in both the Financial toolbox and the Financial Derivatives toolbox is available in the User's Guide available at The MathWorks, Inc. website: www.mathworks.com.

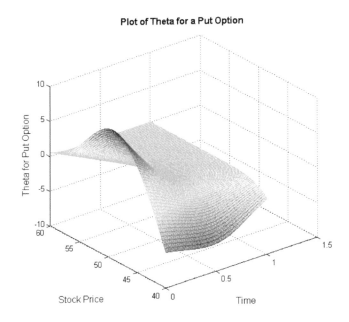

Figure 11.22: Plot of Theta for a put option vs. stock price and time.

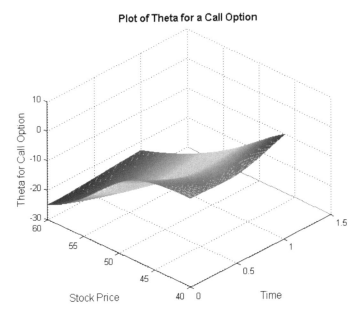

Figure 11.23: Plot of theta for a call option vs. stock price and time.

11.9 References

[1] M. Capinski and T. Zastawniak, Mathematics for Finance: An Introduction to Financial Engineering, Springer-Verlag, New York, 2003.

[2] F. Black and M. Scholes, The Pricing of Options on Corporate Liabilities, J. Political Economic, May-June 1973, pp. 637-654.

[3] J. Hull, Options, Futures and Other Derivatives, Prentice Hall Book Co., Sixth Edition, 2006.

[4] J.C. Cox, S. A. Ross, and M. Rubinstein, Option Pricing: A Simplified Approach, J. Financial Economics, No. 7, Oct. 1979, pp. 229-264.

[5] M. Hoyle, http://www.mathworks.com/matlabcentral/fileexchange/ loadFile.do?objectId =16476.

[6] P. Brandimarte, Numerical Methods in Finance, J. Wiley & Sons, New York, 2003.

[7] F. A. Longstaff and E. S. Schwartz, "Valuing American Options by Simulation: A Simple Least-Squares Approach", May 9, 2001, The Review of Financial Studies, Vol. 14, No. 1, pp. 113-147.

Appendix A

This is the m-file for the finite differences method and it is a modification of Hoyle's m-file given in the MATLAB Central website [5]. This m-file is available at the book website.

% Price via Finite differences
% Solve the Black Scholes PDE by a finite difference approach. We take
% the Black Scholes PDE:
%
% Parameters
% Set up parameters for the option. American option price will depend
% on the current price of the underlying asset _S0_, the strike price _K_,
% the risk free interest rate_r_, the time to maturity of the option _T_ and
% the volatility sigma
S0 = 50; % Initial price
K = 50; % Strike price
r = 0.1; % Interest rate
sigma = 0.4; % Volatility
T = 5/12; % Maturity (years)
% Finite difference parameters
% In addition to the above, to price the option via finite differences, we
% need to specify some extra parameters.
% These are the maximum share price
% that we want to consider in the algorithm (this should be taken to be
% large enough so that the option has effectively 0 value at this price
% throughout its life), and the grid size - the grid being taken in time
% and share price. For the purposes of this file we take
Smax = 2e^(rT)max(S0,K)
%
% Twice the interest adjusted of the maximum of the initial price and the
% strike price at maturity. Along with the grid size we could vary this
% parameter to investigate the effect this choice has on the final value of
% the share price.

```
%
% Given the parameters above now construct the matrices required for
% the finite diffference operation. Here we form D the matrix we need to
% "divide" by as a sparse tridiagonal matrix
N = 50; % number of timesteps
M = 50; % number of Share prices to look at
%[D, a, c, t, dt, s, dS] = SetupFDMatrix(S0, K, r, T, sigma, N, M);
t = linspace(0, T, N+1); dt = T/N;
Smax = 2*max(S0, K); % Maximum price considered
dS = Smax/(M);
s = 0: dS: Smax;
J = (1: M-1)';
a = r/2*J*dt-1/2*sigma^2*J.^2*dt;
b = 1+sigma^J.^2*dt+r*dt;
c = -r/2*J*dt-1/2*sigma^2*J.^2*dt;
D = spdiags([[a(2: end); 0] b [0; c(1: end-1)]],[-1 0 1], M-1, M-1);
% Boundary conditions
P = NaN*ones(N+1, M+1); % Pricing Matrix (t, S)
P(end,:) = max(K-(0: M)*dS, 0); % Value of option at maturity
P(:,1) = K; % Value of option when stock price is 0
P(:, end) = 0; % Value of option when S = Smax
idx = find(s < S0); idx = idx(end); a = S0-s(idx); b = s(idx+1)-S0;
Z = 1/(a+b)*[b a]; % Interpolation vector
% create a Logical index for points close to the initial asset price
%
kdx = (s >= S0-10) & (s <= S0+10);
% Finite difference solver
for ii = N: -1: 1
y = P(ii+1, 2: end-1)'+[-a(1)*K; zeros(M-3, 1); -c(end)*0];
x = D; % Solve FD-equation
P(ii, 2: end-1) = max(x, K-s(2: end-1)');
% Take into account early exercise option
pause(.1);
end
% Price the option
% Extract the final price and plot the resulting surface, and plot a line
% corresponding to the initial asset price.
P_FD = Z*P(1, idx: idx+1)'
% Option price
```

Appendix B

Description of parameters in intenvset

Its general form is

$$[\text{RateSpec}, \text{RateSpecOld}] = \text{intenvset}(\text{RateSpec}, \text{'Argument1'}, ...$$
$$\text{Value1}, \text{'Argument2'}, \text{Value2},...)$$

Each of the arguments is defined as:

RateSpec (Optional) An existing interest-rate specification structure to be changed, probably created from a previous call to intenvset.

The other arguments are given by the name and followed by a value:

CompoundingScalar A value representing the rate at which the input zero rates were compounded when annualized. Default $= 2$. This argument determines the formula for the discount factors:

Compounding $= 1, 2, 3, 4, 6, 12$ Disc $= (1 + Z/F)^{(-T)}$, where F is the compounding frequency, Z is the zero rate, and T is the time in periodic units; for example, $T = F$ is 1 year.

Compounding $= 365$ Disc $= (1 + Z/F)^{(-T)}$, where F is the number of days in the basis year and T is a number of days elapsed computed by basis.

Compounding $= -1$ Disc $= \exp(-T*Z)$, where T is time in years.

Disc The number of points (NPOINTS) by number of curves (NCURVES) matrix of unit bond prices over investment intervals from StartDates, when the cash flow is valued, to EndDates, when the cash flow is received.

Rates The number of points (NPOINTS) by number of curves (NCURVES) matrix of rates in decimal form. For example, 5% is 0.005 in Rates. Rates are the yields over investment intervals from StartDates, when the cash flow is valued, to EndDates, when the cash flow is received.

EndDates An NPOINTS-by-1 vector or scalar of serial maturity dates ending the interval to discount over.

StartDates An NPOINTS-by-1 vector or scalar of serial dates starting the interval to discount over. Default = ValuationDate.

ValuationDate (Optional) A scalar value in serial date number form representing the observation date of the investment horizons entered in StartDates and EndDates. Default = min(StartDates).

Basis (Optional) The Day-count basis of the instrument. A vector of integers. If it is a 0 = actual/actual (default). If it is a 1 = 30/360 (SIA). If it is a 2 = actual/360. If it is a 3 = actual/365. If it is a 4 = 30/360 (PSA). If it is a 5 = 30/360 (ISDA). If it is a 6 = 30/360 (European). If it is a 7 = actual/365 (Japanese). If it is an 8 = actual/actual (ISMA). If it is a 9 = actual/360 (ISMA). If it is a 10 = actual/365 (ISMA). If it is an 11 = 30/360E (ISMA)

EndMonthRule (Optional) The End-of-month rule. It is a vector. This rule applies only when Maturity is an end-of-month date for a month having 30 or fewer days. If set to 0 = ignore rule, meaning that a bond's coupon payment date is always the same numerical day of the month. 1 = set rule on (default), meaning that a bond's coupon payment date is always the last actual day of the month.

Index

T - #0343 - 071024 - C48 - 234/156/19 - PB - 9780367384982 - Gloss Lamination